Global Well-posedness and Asymptotic Behavior of the Solutions to Non-classical Thermo(visco)elastic Models

Yuming Qin · Zhiyong Ma

Global Well-posedness and Asymptotic Behavior of the Solutions to Non-classical Thermo(visco)elastic Models

Springer

Yuming Qin
Department of Applied Mathematics
Donghua University
Shanghai
China

Zhiyong Ma
Shanghai Second Polytechnic University
Shanghai
China

ISBN 978-981-10-1713-1 ISBN 978-981-10-1714-8 (eBook)
DOI 10.1007/978-981-10-1714-8

Library of Congress Control Number: 2016943327

Mathematics Subject Classification (2010): 35B40, 80M35

Printed on acid-free paper

This Springer imprint is published by Springer Nature
The registered company is Springer Science+Business Media Singapore Pte Ltd.

To our parents
To Yu Yin and Jia Qin
To Zhangmi Huang

Preface

This book is designed to present recent results on some global well-posedness and asymptotic behavior of the solutions to non-classical thermo(visco)elastic models arising from physics, mechanics, and materials science such as thermoviscoelastic systems, thermoelastic systems of types II, III, and with second sound, Timoshenko-type system with past history. Some of the content of this book is based on the research carried out by the authors and their collaborators in recent years. Most of it has been previously published only in original papers, and some of the material has never been published until now. Therefore, the authors hope that the book will benefit both the interested beginner in the field and the expert.

This book is divided into nine chapters.

Chapter 1 is devoted to introducing some basic knowledge in modern analysis, some useful inequalities and basic theory of semigroups.

In Chap. 2, we investigate one-dimensional nonlinear thermoelasticity with thermal memory and second sound. Thermoelastic model is based on the continuum physics and thermodynamics, considering the distribution of the thermoelastic body deformation and temperature.

The classical model for the heat propagation turns into the well-known Fourier law

$$q + k\nabla\theta = 0, \tag{1}$$

where θ is temperature (difference to a fixed constant reference temperature), q is the heat conduction vector and k is the coefficient of thermal conductivity. The model using classic Fourier law exhibits the physical paradox of infinite propagation speed of signals. To eliminate this paradox a generalized thermoelasticity theory has been developed subsequently. The development of this theory was accelerated by the advent of the second sound effects observed experimentally in materials at a very low temperature. In heat transfer problems involving very short time intervals and/or very high heat fluxes, it has been revealed that the inclusion of the second sound effects to the original theory yields results which are realistic and very much different from those obtained with the classical Fourier's law.

The first theory, developed by Lord and Shulman [156], replaces (1) with the Cattaneo–Maxwell law

$$\tau_0 q_t + q + k\theta_x = 0. \tag{2}$$

The heat equation associated with (2) becomes hyperbolic and, hence, automatically eliminates the paradox of infinite speeds. The positive parameter τ_0 is the relaxation time describing the time lag in the response of the heat flux to a gradient in the temperature. In this chapter, we shall prove the global existence and exponential stability of solutions to nonlinear thermoelastic equations with second sound provided that the initial data are close to the equilibrium and the relaxation kernel is strongly positive definite and decays exponentially.

In Chap. 3, we consider the Timoshenko-type system with a past history. The model of a homogeneous Timoshenko beam is described by a system of second-order partial differential equations. Timoshenko [270] first proposed the transverse vibrations of a beam which are given by the following coupled partial differential equations

$$\begin{cases} \rho u_{tt} = (k(u_x - \varphi))_x, & \text{in} \quad (0,L) \times (0, +\infty), \\ I_\rho \varphi_{tt} = (EI\varphi_x)_x + k(u_x - \varphi), & \text{in} \quad (0,L) \times (0, +\infty), \end{cases}$$

where t denotes the time variable and x is the space variable along the beam, the length of which is L, in its equilibrium configuration. The function u is the transverse displacement of the beam and φ is the rotation angle of the filament of the beam. The coefficients ρ, I_ρ, E, I and k are, respectively, the mass per unit length, the polar moment of the inertia of a cross section, Young's modulus of elasticity, the moment of inertia of a cross section, and the shear modulus. The model considered here is a modified Timoshenko system, that is, a Timoshenko system in classical thermoelasticity of type I. We shall use multiplier techniques to prove the stability property for the system with a past history. For a kernel of polynomial decay, we prove the polynomial stability results for the equal wave-speed propagation, and establish a decay result for the nonequal wave-speed case under the assumption that g decays exponentially. Moreover, the existence of the global attractor is achieved.

In Chap. 4, as a continuation of Chap. 3, we consider a modified Timoshenko system, that is, a Timoshenko system in classical thermoelasticity of type III, we shall use the semigroup method to establish a polynomial stability result for Timoshenko-type system with a past history for the nonequal wave-speed case.

In Chap. 5, we study the (non)linear (non)homogeneous thermoelastic Bresse system. In their study on networks of flexible beams, Lagnese, Leugering, and Schmidt [9] derived a general model for three-dimensional nonlinear thermoelastic beams. A special case of this model is a linear planar, shearable thermoelastic beam whose motion is governed by the following system of partial differential equations

$$
\begin{cases}
\rho h w_{1tt} = (Eh(w_1' - kw_3) - \alpha\theta_1)' - kGh(\phi_2 + w_3' + kw_1), \\
\rho h w_{3tt} = Gh(\phi_2 + w_3' + kw_1)' + kEh(w_1' - kw_3) - k\alpha\theta_1, \\
\rho I \phi_{2tt} = EI\phi_2'' - Gh(\phi_2 + w_3' + kw_1) - \alpha\theta_3', \\
\rho c\theta_{1t} = \theta_1'' - \alpha T_0(w_{1t}' - kw_{3t}), \\
\rho c\theta_{3t} = \theta_3'' -_0 \phi_{2t}',
\end{cases} \tag{3}
$$

where w_1, w_2 and ϕ_2 are the longitudinal vertical and shear angle displacements; θ_1, θ_3 are the temperature deviations from the reference temperature T_0 along the longitudinal and vertical directions; $E, G, \rho, I, m, h, k, c$ are positive constants for the elastic and thermal material properties. We shall use the semigroup method and establish the global existence of solutions for the thermoelastic Bresse system.

In Chap. 6, as a continuation of Chap. 5, we shall consider the stability for thermoelastic bresse system, that is, we use multiplier techniques to prove the exponential stability result only for the equal wave-speed case.

In Chap. 7, we consider the linear thermoelastic model of type III with memory effects. In 1990s, Green and Naghdi [89–92] introduced three new types of thermoelastic theories in the aim of replacing the usual entropy production inequality with an entropy balance law. In each of these theories, the heat flux is given by a different constitutive assumption. As a result, three theories were obtained and respectively called thermoelasticity type I, type II, and type III. When the theory of type I is linearized, we obtain the classical system of thermoelasticity. The systems arising in thermoelasticity of type III are of dissipative nature whereas those of type II thermoelasticity do not sustain energy dissipation. In this chapter, we shall establish the global existence result for the higher-dimensional linear thermoviscoelastic equations of type III by using a semigroup approach. Using the multiplier techniques and Lyapunov methods, we prove that the energy for such a model decays to zero exponentially by introducing a velocity feedback on a part of the boundary of a thermoelastic body, which is clamped along the rest of its boundary to increase the loss of energy.

In Chap. 8, we study the thermoelastic model of type II. The issue of the asymptotic behavior of thermoelastic systems has attracted much attention in recent years. In the beginning, the mathematicians and engineers mainly considered the behavior of thermoelastic systems under the theory of classical thermoelasticity, in which the heat flux is given by the Fourier's law. Note that the theory of classical thermoelasticity predicts an infinite speed of heat propagation. This leads to an unrealistic property that a sudden disturbance at some point will be felt instantly everywhere else in the materials. In 1990s, Green and Naghdi [89–92] proposed three types of thermoelastic theories called thermoelasticity of types I, II, III, respectively, based on an entropy equality instead of the usual entropy inequality. The thermoelasticity of type I coincides with the classical one. In type II, known as thermoelasticity without dissipation, the heat is allowed to propagate by means of thermal waves, but without dissipating. The types I and II are limiting cases of thermoelasticity type III. Based on these three new theories of thermoelasticity, many engineers and mathematicians discussed the asymptotic behavior of several thermoelastic problems so as to describe the thermo-mechanical interactions in elastic materials. In this chapter, we shall prove the global existence for the three-dimensional thermoelastic equations of type II by means of semigroup methods.

In Chap. 9, we shall consider the thermoviscoelastic system.

For very viscous liquids, the ordinary hydrodynamic description needs to be generalized to allow for the slow relaxation processes related to the high viscosity. As far as mechanical relaxation is concerned, the theory of viscoelasticity provides such a description. However, the slow structural relaxation, which causes the high viscosity, also leads to a slow relaxation of thermal variables like temperature or entropy. Effects of such thermal relaxation are a frequency dependence of the specific heat and a coupling of heat conduction with structural relaxation. For a complete generalization of hydrodynamics for very viscous liquids, therefore, the relaxation of mechanical and thermal variables has to be treated on the same level. The result of this generalization will be a theory of linear thermoviscoelasticity. In this chapter, we shall obtain a decay result for higher-dimensional linear thermoviscoelastic equations by introducing a velocity feedback on a part of the boundary and using the multiplier technique method.

We sincerely wish that the reader will know the main ideas and essence of the basic theories and methods in deriving the global existence, uniqueness, asymptotic behavior of solutions for the models considered in this book. We also wish that the reader can be stimulated by some ideas from this book and undertake the further study after having read the related references and bibliographic comments in this book.

We also want to take this opportunity to thank all the people who concern about us including our teachers, colleagues, and collaborators. Yuming Qin also acknowledges the NNSF of China for its support. Currently, this book project is being supported by the NNSF of China with contract no. 11271066 and, by the grant from Shanghai Education Commission with contract no. 13ZZ048.

Last but not least, we want to express our deepest thanks to our parents, sisters, brothers, and wives for their great help, constant concern, and advice in our career.

Shanghai, China

Yuming Qin
Zhiyong Ma

Contents

Chapter 1
Preliminary

In this chapter, we recall some basic knowledge in functional analysis and harmonic analysis, most of which will be used in the subsequent chapters. The reader can easily find the detailed proofs in the related literature, see, e.g., Adams [1], Bahouri, Chemin and Danchin [10], Brezis [22], Chueshov [39], Courant and Hilbert [46], Dafermos [49], DiPerna and Lions [58], Evans [64, 65], Feireisl [77], Gilbarg and Trudinger [85], Hale [103], Hardy, Littlewood and Polya [107], Henry [109], Hille and Phillips [111], Jiang and Racke [122], Kato [124], Lieb and Loss [144], Liu and Zheng [155], Liu and Zheng [153], Miao [181, 182], Miao, Nirenberg [198], Pazy [201], Qin [210, 214, 215], Rudin [246], Schwartz [248], Serrin [249], Sobolev [258], Taylor [268], Walter [273], Yosida [283], Zheng [286, 287].

1.1 Some Definitions and Theorems in Modern Analysis

In this book, we denote by $\mathbb{R}^n$ the Euclid space, $\Omega \subset \mathbb{R}^n$ is a bounded smooth domain, $C^k(\mathbb{R}^n)$ $(or\ C^k(\Omega))$ is the kth differentiable continuous functions space in $\mathbb{R}^n$ $(or\ \Omega)$, $C^\infty(\mathbb{R}^n)$ $(or\ C^\infty(\Omega))$ is the ∞th differentiable continuous functions space in $\mathbb{R}^n$ $(or\ \Omega)$, $C_c^\infty(\mathbb{R}^n)$ $(or\ C_c^\infty(\Omega))$ is the ∞th differentiable continuous functions space with compact support in $\mathbb{R}^n$ $(or\ \Omega)$. The norm for the space $C^k(\mathbb{R}^n)$ is defined as

$$C^k(\mathbb{R}^n) = \left\{ f \middle| f \in C^k(\mathbb{R}^n),\ \|f\|_{C^k(\mathbb{R}^n)} = \sum_{|\alpha| \le k} \sup_{x \in \mathbb{R}^n} |\partial^\alpha f(x)| < +\infty \right\}, \tag{1.1.1}$$

where $\partial^\alpha f = \frac{\partial^{|\alpha|} f}{\partial x_1^{\alpha_1} \cdots \partial x_n^{\alpha_n}}$, $\alpha = (\alpha_1, \alpha_2, \ldots, \alpha_n)$, $|\alpha| = \alpha_1 + \alpha_2 + \cdots + \alpha_n$.

Y. Qin and Z. Ma, *Global Well-posedness and Asymptotic Behavior of the Solutions to Non-classical Thermo(visco)elastic Models*,
DOI 10.1007/978-981-10-1714-8_1

The notation $\mathcal{S}(\mathbb{R}^n)$ denotes the Schwartz space defined as

$$\mathcal{S}(\mathbb{R}^n) = \left\{\phi \middle| \phi \in C^\infty(\mathbb{R}^n),\ \|\phi\|_{(\alpha,\beta)} = \sup_{x\in\mathbb{R}^n} |x^\alpha \partial^\beta \phi(x)| < +\infty,\ for\ all\ \alpha, \beta \in \mathbb{N}^n\right\}. \tag{1.1.2}$$

The space $\mathcal{S}(\mathbb{R}^n)$ endowed the quasi-norm $\|\cdot\|_{(\alpha,\beta)}$ in (1.1.2) is a Fréchet space, and $C_c^\infty(\mathbb{R}^n) \subset \mathcal{S}(\mathbb{R}^n) \subset C^\infty(\mathbb{R}^n)$, the space $\mathcal{S}'(\mathbb{R}^n)$ is the dual space of $\mathcal{S}(\mathbb{R}^n)$. $\mathcal{S}(\mathbb{R}^n)$ is the Schwartz class of rapidly decreasing functions.

Let (X, M, μ) be a non-atomic measurable space, for a complex-or real-valued measurable function $f(x)$ defined on X, its distributional function is defined by

$$f_*(\sigma) = \mu\left\{x \in X : f(x) > \sigma\right\},\ \text{for all } \sigma > 0, \tag{1.1.3}$$

which is non-increasing and continuous from the right. Furthermore, its non-increasing rearrangement f^* is defined by

$$f^*(t) = \inf\left\{s > 0 : f_*(s) \le t\right\},\ \text{for all } t > 0, \tag{1.1.4}$$

which is also non-increasing and continuous from the right and has the same distributional function as $f(x)$. $\mathcal{E}'(\mathbb{R}^3)$ is the space of all tempered distributions on $\mathbb{R}^3$, $\mathcal{D}(\mathbb{R}^3)$ is the set of all scalar polynomials defined on $\mathbb{R}^3$.

$L^p(\mathbb{R}^3)$ $(1 \le p \le +\infty)$ denotes the Lebesgue space of all L^p integrable functions associated with the norms

$$\|f\|_p = \left(\int_{\mathbb{R}^3} |f(x)|^p dx\right)^{1/p},\quad 1 \le p < +\infty, \tag{1.1.5}$$

$$\|f\|_\infty = ess \sup_{x\in\mathbb{R}^3} |f(x)|,\quad p = +\infty. \tag{1.1.6}$$

Definition 1.1.1 Let $\Omega \subset \mathbb{R}^n$ be an open set, $\mathcal{D}(\Omega)$ is the set of all ϕ in $C_0^\infty(\Omega)$ endowed with a topology so that $\{\phi_i\}$ converges to an element ϕ in $\mathcal{D}(\Omega)$ if and only if

(1) there exists a compact set $K \subset \Omega$ such that $\phi_i \subset K$ for every i;
(2) $\lim_{i\to+\infty} D^\alpha \phi_i = D^\alpha \phi$ uniformly on K for each multi-index α.

Definition 1.1.2 Let $d_{K,m}(\Omega) = \sup_{x\in K} \sum_{|\alpha|=m} |D^\alpha \phi(x)|$, where K is compact in Ω. $\mathcal{D}'(\Omega)$ is the dual space of $\mathcal{D}(\Omega)$, and

$$\langle D^\alpha T, \phi\rangle = (-1)^{|\alpha|}\langle T, D^\alpha \phi\rangle,\quad for\ \ all\ \phi \in \mathcal{D}(\Omega), \tag{1.1.7}$$

where

$$D^{\alpha}T = \frac{\partial^{\alpha_1} \cdots \partial^{\alpha_n}}{\partial x_1^{\alpha_1} \cdots \partial x_n^{\alpha_n}} \in \mathcal{D}(\Omega), \tag{1.1.8}$$

with $|\alpha| = \sum_{i=1}^{n} \alpha_i$.

Definition 1.1.3 $f \in L^1_{loc}(\Omega)$ defines a distribution $T_f \in \mathcal{D}'(\Omega)$ by

$$\langle T_f, \phi \rangle = Re\left(\int_{\Omega} f(x)\overline{\phi(x)}dx\right), \quad for \ all \ \phi \in \mathcal{D}(\Omega). \tag{1.1.9}$$

Definition 1.1.4 The Sobolev space $W^{m,p}(\Omega)$ is defined as

$$W^{m,p}(\Omega) = \left\{u \in L^p(\Omega)\middle| D^{\alpha}u \in L^p(\Omega), \ |\alpha| \leq m\right\}, \tag{1.1.10}$$

where $0 < m, p \leq +\infty, \ p, \ m \in \mathbb{N}$.

If $u \in W^{m,p}(\Omega)$, then the norm of u is defined as

$$\|u\|_{W^{m,p}(\Omega)} = \left(\sum_{0\leq|\alpha|\leq m} \|D^{\alpha}u\|^p_{L^p(\Omega)}\right)^{\frac{1}{p}}. \tag{1.1.11}$$

Specially, when $m = 0$, $W^{0,p} = L^p$; when $p = 2$, $W^{m,2}(\Omega)$ is denoted as $H^m(\Omega)$, the norm of $u \in H^m(\Omega)$ is defined as

$$\|u\|_{H^m(\Omega)} = \left(\sum_{0\leq|\alpha|\leq m} \int_{\Omega} \left|D^{\alpha}u(x)\right|^2 dx\right)^{1/2} \tag{1.1.12}$$

and the inner product is expressed as

$$(u, v)_{H^m(\Omega)} = \sum_{0\leq|\alpha|\leq m} Re \int_{\Omega} D^{\alpha}u(x)\overline{D^{\alpha}v(x)}dx. \tag{1.1.13}$$

Theorem 1.1.5 (Adams [1]) *The continuous embeddings in Sobolev space are as:*

(1) When $1 \leq p < n$, $\frac{1}{p} - \frac{m}{n} > 0$, *we have* $W^{m,p}(\Omega) \hookrightarrow L^q(\Omega)$, *where* $q \in [p, \frac{np}{n-p}]$.
(2) If $p = n > 1$, *then* $W^{m,p}(\Omega) \hookrightarrow L^q(\Omega)$ *for all* $q \in [p, +\infty)$.
(3) When $p = n = 1$, *there holds* $W^{1,1}(\Omega) \hookrightarrow L^{\infty}(\Omega)$.
(4) When $p = 1$ *or* $p = +\infty$, *the Sobolev spaces* $W^{k,1}(\Omega)$ *and* $W^{k,\infty}(\Omega)$ *do not have the reflexible property.*

(5) *If* $\frac{1}{p} - \frac{m}{n} < 0$, *then* $W^{m,p}(\Omega) \hookrightarrow C^{k,\alpha}(\Omega)$, *here* $k = [m - \frac{n}{p}]$, $\alpha = \left(m - \frac{n}{p}\right) = m - \frac{n}{p} - k$.

Here $[\cdot]$ and $(\cdot)$ denote the integral part and fractional part of numbers respectively.

Definition 1.1.6 The Sobolev–Slobodeckiî space $W^{k+\beta,p}(Q)$ (Q is a domain in $\mathbb{R}^L$, $1 \le p < +\infty$, $0 < \beta < 1$, $k = 0, 1, 2, \ldots$) is defined as

$$W^{k+\beta,p}(Q) = \left\{ u \in L^p(\Omega) \middle| \|u\|_{W^{k+\beta,p}(Q)} = \left(\|u\|_{W^{k,p}(Q)} + \sum_{|\alpha|=k} \int_Q \int_Q \frac{|\partial^\alpha v(y) - \partial^\alpha v(z)|^p}{|y-z|^{L+\beta p}} dydz \right)^{\frac{1}{p}} \right\}. \tag{1.1.14}$$

Remark 1.1.1 Here we introduce some examples of special Sobolev–Slobodeckiî spaces. The topology of homogeneous space $\dot{W}_2^l(\Omega)$ is equipped as

$$\|u\|^2_{\dot{W}_2^l(\Omega)} = \begin{cases} \sum_{|\alpha|=l} \|\partial_x^\alpha u\|^2_{L^2(\Omega)}, & \text{if } l \in \mathbb{Z}, \\ \sum_{|\alpha|=[l]} \int_\Omega \int_\Omega \frac{|\partial^\alpha u(x) - \partial^\alpha u(y)|^2}{|x-y|^{m+2(l-[l])}} dxdy, & \text{if } l \in \mathbb{R}^+/\mathbb{Z}^+, \end{cases} \tag{1.1.15}$$

where $[l]$ is the integer part of $l \in \mathbb{R}^+$, $\Omega \subset \mathbb{R}^n$ is a smooth domain.

The topology of non-homogeneous space $W_2^l(\Omega)$ is equipped as

$$\|u\|_{W_2^l(\Omega)} = \left(\sum_{|\alpha| \le l} \|\partial_x^\alpha u\|^2_{L^2(\Omega)} + \|u\|^2_{\dot{W}_2^l(\Omega)} \right)^{\frac{1}{2}}. \tag{1.1.16}$$

Denoting $Q_T = \Omega \times (0, T)$, we can define the space $W_2^{l,\frac{l}{2}}(Q_T)$ as

$$W_2^{l,\frac{l}{2}}(Q_T) = L^2(0, T; W_2^l(Q)) \cap L^2(Q; W_2^{\frac{l}{2}}(0, T)) \tag{1.1.17}$$

with the norm

$$\begin{aligned} \|u\|_{W_2^{l,\frac{l}{2}}(Q_T)} &= \left(\|u\|^2_{W_2^{l,0}(Q_T)} + \|u\|^2_{W_2^{0,\frac{l}{2}}(Q_T)} \right)^{\frac{1}{2}} \\ &= \left(\int_0^T \|u\|^2_{W_2^l(Q)} dt + \int_\Omega \|u\|^2_{W_2^{\frac{l}{2}}(0,T)} dx \right)^{\frac{1}{2}}, \end{aligned} \tag{1.1.18}$$

where the topology equipped for the space $W_2^l(0,T)$ is defined as

$$\|u\|_{W_2^l(0,T)} = \begin{cases} \left(\sum_{j=0}^{l} \|D_t^j u\|^2_{L^2(0,T)}\right)^{\frac{1}{2}}, \text{ if } l\in\mathbb{Z}, \\ \left(\sum_{j=0}^{l} \|D_t^j u\|^2_{L^2(0,T)} + \int_0^T \int_0^t \frac{|D_t^{[l]}u(t) - D_\tau^{[l]}u(\tau)|^2}{|t-\tau|^{l+2(l-[l])}} d\tau dt\right)^{\frac{1}{2}}, \text{ if } l \in \mathbb{R}^+/\mathbb{Z}^+. \end{cases} \tag{1.1.19}$$

1.2 Some Useful Inequalities

In this section, we shall recall some inequalities which will be used in the subsequent chapters.

Theorem 1.2.1 (Young [284]) (Young's Inequality) *The following inequalities hold*

$$ab \le \frac{a^p}{p} + \frac{b^q}{q}, \quad q = \frac{p}{p-1}, \quad 1 < p < +\infty, \quad \textit{for all } a, b > 0, \tag{1.2.1}$$

$$ab \le \frac{\varepsilon}{p} a^p + \frac{1}{q\varepsilon^{\frac{1}{p-1}}} b^q, \quad q = \frac{p}{p-1}, \quad 1 < p < +\infty, \quad \textit{for all } a, b, \varepsilon > 0. \tag{1.2.2}$$

Theorem 1.2.2 (Adams [1]) (The Cauchy–Schwarz Inequality) *There holds that*

$$|x \cdot y| \le |x||y|, \textit{ for all } x, \ y \in \mathbb{R}^n, \tag{1.2.3}$$

here $|x| = (x, x)^{1/2} = (\Sigma_{i=1}^n x_i^2)^{1/2}$ *for all* $x \in \mathbb{R}^n$.

Theorem 1.2.3 (Hölder [113]) (Hölder's Inequality) *Let* $\Omega \subseteq \mathbb{R}^n$ *be a domain, assume that* $u \in L^p(\Omega)$, $v \in L^q(\Omega)$ *with* $1 \le p, q \le +\infty$ *and* $\frac{1}{p} + \frac{1}{q} = 1$. *Then*

$$\int_\Omega |uv| dx \le \|u\|_{L^p(\Omega)} \|v\|_{L^q(\Omega)}. \tag{1.2.4}$$

In 1896, Minkowski established the following famous inequality.

Theorem 1.2.4 (Adams [113]) (Minkowski's Inequality) *Assume* $1 \le p \le +\infty$. *Then for any* $u, v \in L^p(\Omega)$,

$$\|u + v\|_{L^p(\Omega)} \le \|u\|_{L^p(\Omega)} + \|v\|_{L^p(\Omega)}. \tag{1.2.5}$$

Theorem 1.2.5 (Cazenave [32]) (Poincaré's Inequality) *Assume that* $\Omega \subset \mathbb{R}^n$ *is a bounded smooth domain, then there holds*

$$\|u\| \leq \lambda_1^{-\frac{1}{2}} \|u\|_1, \quad \textit{for all } u \in H_0^1(\Omega), \tag{1.2.6}$$

where λ_1 *is the first eigenvalue of* A *under the homogeneous Dirichlet boundary condition,* $\|u\|_1$ *is the norm of* u *in* H_0^1, $A = -\Delta$.

Theorem 1.2.6 (Gronwall [97]) (Gronwall's Inequality) *Let* $a(t) \in L^1(0, T)$, $a \geq 0$, $\beta(t) \in L^1(0, T)$, $b_0 \in \mathbb{R}$, $b(\tau) = b_0 + \int_0^\tau \beta(t)dt$. *Assume* $\gamma(t) \in L^\infty(0, T)$ *satisfies*

$$\gamma(\tau) \leq b(\tau) + \int_0^\tau a(t)\gamma(t)dt \ \textit{ for almost all (a.a.) } \tau \in [0, T]. \tag{1.2.7}$$

Then for a. a. $\tau \in [0, T]$, *we have*

$$\gamma(\tau) \leq b_0 \exp\left(\int_0^\tau a(s)ds\right) + \int_0^\tau \beta(t) \exp\left(\int_t^\tau a(s)ds\right) dt. \tag{1.2.8}$$

Under assumptions of Theorem 1.2.6, the following inequalities hold for dimension $n = 3$.

Theorem 1.2.7 (Cazenave [32]) (The Gagliardo-Nirenberg Inequality)

$$\|u\|_{L^{\frac{6}{3-2\varepsilon}}} \leq C\|u\|^{1-\varepsilon} \|u\|_1^\varepsilon, \quad 0 \leq \varepsilon \leq 1, \quad \textit{for all } u \in H_0^1(\Omega), \tag{1.2.9}$$

$$\|u\|_{L^p} \leq C\|u\|^{\frac{2}{p}} \|u\|_{\frac{3}{2}}^{1-\frac{2}{p}}, \quad p \in [2, +\infty), \quad \textit{for all } u \in V_{\frac{3}{2}}. \tag{1.2.10}$$

Theorem 1.2.8 (Martinez [163]) *Let* $\phi(t) : \mathbb{R}^+ \to \mathbb{R}^+$ *be a non-increasing function and* $\sigma : \mathbb{R}^+ \to \mathbb{R}^+$ *be a strictly increasing* C^1 *function, with* $\sigma(t) \to +\infty$ *as* $t \to +\infty$. *Assume that there exist* $p, q \geq 0$ *and* $c > 0$ *such that*

$$\int_s^{+\infty} \sigma'(t)\phi^{1+p}(t)dt \leq c\phi^{1+p}(s) + \frac{c\phi(s)}{\sigma^q(s)}, \quad 1 \leq s < +\infty.$$

Then there exist positive constants k *and* δ *such that*

$$\begin{cases} \phi(t) \leq k e^{-\delta\sigma(t)}, & \textit{for all } t \geq 1, \text{ if } \ p = q = 0, \\ \phi(t) \leq \frac{k}{\sigma(t)^{(1+q)/p}}, & \textit{for all } t \geq 1, \text{ if } \ p > 0. \end{cases}$$

1.3 Basic Theory of Semigroups

In this section, we recall some basic knowledge in semigroups, most of which will be used in the subsequent chapters. The reader can easily find the detailed proofs in the related literature, see, e.g., Batty and Duyckaerts [13], Belleni-Morante and McBride [14], Borichev and Tomilov [19], Liu and Zheng [155], Prüss [205], Qin [210], Pazy [201].

Definition 1.3.1 (*Pazy [201], Zheng [287]*) (*Semigroups*) Let X be a Banach space or a closed subset of a Banach space, the parameter family $S(t),\ 0 \le t \le +\infty$ from X to X is called a semigroup if

(i) $S(0) = Id$ (Identity operator on X),
(ii) $S(t+s) = S(t)S(s)$.

Definition 1.3.2 (*Pazy [201], Zheng [287]*) (*C_0-Semigroups*) A semigroup $S(t)$, $0 \le t \le +\infty$ from X to X is called a strong continuous semigroup of bounded linear operators if

$$\lim_{t\to 0^+} S(t)x = x \ or \ \lim_{t\to 0^+} \|S(t)x - x\| = 0,\ for\ all\ x \in X, \tag{1.3.1}$$

i.e., $T(t)$ is a C_0-semigroup.

Definition 1.3.3 (*Pazy [201], Zheng [287]*) (*Contraction Semigroups*) The semigroup $S(t)$ is a contraction semigroup if there exists a constant $\alpha > 0$ $(0 < \alpha < 1)$ such that for all $t > 0$,

$$\|S(t)x - S(t)y\| \le \alpha \|x - y\|,\ for\ all\ x,\ y \in X. \tag{1.3.2}$$

Definition 1.3.4 (*Pazy [201], Zheng [287]*) (*Analytic Semigroups*) Let $\Delta = \Big\{z|\varphi_1 < argz < \varphi_2,\ \varphi_1 < 0 < \varphi_2\Big\}$ and for all $z \in \Delta$, $S(z)$ be a bounded linear operator. The family $S(z)$ $(z \in \Delta)$ is an analytic semigroup in Δ if

(i) $z \mapsto S(z)$ is analytic in Δ,
(ii) $S(0) = Id$, i.e., $\lim_{z\to 0} S(z)x = x$ for every $x \in X$,
(iii) $S(z_1 + z_2) = S(z_1)S(z_2)$ for all $z_1,\ z_2 \in \Delta$. A semigroup $S(t)$ is called analytic if it is analytic in some sector containing the nonnegative real axis.

Definition 1.3.5 (*Pazy [201], Zheng [287]*) The linear operator $\mathcal{A}$ defined by

$$D(\mathcal{A}) = \{x \in X : \lim_{t\to 0^+} (S(t)x - x)/t\ exists\} \tag{1.3.3}$$

and

$$\mathcal{A}x = \lim_{t\to 0^+} (S(t)x - x)/t = \frac{d(S(t)x)}{dt}|_{t=0}\ for\ all\ x \in D(\mathcal{A}) \tag{1.3.4}$$

is called the infinitesimal generator of the semigroup $S(t)$, $D(\mathcal{A})$ is called the domain of $\mathcal{A}$.

In the following, for the sake of convenience, we introduce the definition of maximal accretive operator (see, e.g., Zheng [287] and Pazy [201]).

Definition 1.3.6 *(Pazy [201], Zheng [287])* Let $\mathcal{A}$ be a linear operator defined in a Banach space X, $\mathcal{A}: D(\mathcal{A}) \subset X \mapsto X$. If for any $x, \ y \in D(\mathcal{A})$ and any $\lambda > 0$,

$$\|x - y\| \leq \|x - y + \lambda(\mathcal{A}x - \mathcal{A}y)\|, \tag{1.3.5}$$

then $\mathcal{A}$ is said to be an accretive operator. Moreover, if $\mathcal{A}$ is densely defined accretive operator, and $I + \mathcal{A}$ is surjective, i.e., $R(I + \mathcal{A}) = X$, then $\mathcal{A}$ is said to be a maximal accretive operator.

Definition 1.3.7 For real numbers $M > 0$ and $\omega \geq 0$, let $\mathcal{G}(M, \omega; X)$ denote the set of generators of C_0-semigroups $S(t), 0 \leq t < +\infty$ on a Banach space X satisfying (1.3.5).

Theorem 1.3.8 *(Pazy [201], Zheng [287]) Let $S(t), 0 \leq t < +\infty$ be a C_0 semigroup on a Banach space X. Then there exist constants $M > 0$ and $\omega \geq 0$ such that*

$$\| S(t) \| \leq Me^{\omega t} \quad \textit{for all } t \geq 0. \tag{1.3.6}$$

Obviously, if $M = 1$ and $\omega = 0$ in (1.3.6), then we obtain a C_0-semigroup of nonexpansions or contractions.

Theorem 1.3.9 *(Pazy [201], Zheng [287]) For any $x \in D, S(t)x \in C^1([0, +\infty), X)$. Moreover, for all $t \geq 0$,*

$$x - S(t)x = \int_0^t \mathcal{A}S(\tau)xd\tau = \int_0^t S(\tau)\mathcal{A}xd\tau, \tag{1.3.7}$$

and

$$\frac{d(S(t)x)}{dt} + \mathcal{A}(S(t)x) = 0. \tag{1.3.8}$$

Theorem 1.3.10 *(Pazy [201], Zheng [287])* (Hille–Yoshida Theorem) $\mathcal{A} \in \mathcal{G}(M, \omega; X)$ *if and only if*

(i) $\mathcal{A}$ is a closed linear operator whose domain $D(\mathcal{A})$ is dense in X; and
(ii) for all real numbers $\lambda > \omega, \lambda \in \rho(\mathcal{A})$ (the resolvent set of $\mathcal{A}$), and

$$\| [R(\lambda, \mathcal{A})]^n \| \leq \frac{M}{(\lambda - \omega)^n} \quad \textit{for } n = 1, 2, \ldots \tag{1.3.9}$$

Theorem 1.3.11 *(Batty [12]) Let $S(t), t > 0$ be a bounded C_0-semigroup on a Banach space X with generator $\mathcal{A}$. Suppose that $i\mathbb{R}$ is contained in the resolvent set $\rho(\mathcal{A})$ of $\mathcal{A}$. Then*

$$\| S(t)\mathcal{A}^{-1} \| \to 0, \quad t \to +\infty. \tag{1.3.10}$$

Lemma 1.3.12 *(Pazy [201], Zheng [287]) Let $\mathcal{A}$ be a linear operator defined in a Hilbert space H, $\mathcal{A} : D(\mathcal{A}) \subset H \to H$. Then the necessary and sufficient conditions for $\mathcal{A}$ being maximal accretive are:*

(i) $\Re(\mathcal{A}x, x) \geq 0$, *for allx* $\in D(\mathcal{A})$,
(ii) $R(I + \mathcal{A}) = H$.

Proof We first prove the necessity. By (1.3.5),

$$(x, x) = \| x \|^2 \leq \| x + \lambda \mathcal{A}x \|^2 = (x, x) + 2\lambda Re(\mathcal{A}, x) + \lambda^2 \| \mathcal{A}x \|^2 . \tag{1.3.11}$$

Thus, for all $\lambda > 0$,

$$Re(\mathcal{A}x, x) \geq -\frac{\lambda}{2} \| \mathcal{A}x \|^2 . \tag{1.3.12}$$

Letting $\lambda \to 0$, we get (i). Furthermore, (ii) immediately follows from the fact that $\mathcal{A}$ is m-accretive.

We now prove the sufficiency. It follows from (i) that for all $\lambda > 0$,

$$\| x - y \|^2 \leq Re(x - y, x - y + \lambda \mathcal{A}(x - y)) \leq \| x - y \| \| x - y + \lambda(\mathcal{A}x - \mathcal{A}y) \|, \tag{1.3.13}$$

i.e., (1.3.5) holds. Now it remains to prove that $\mathcal{A}$ is densely defined. We use a contradiction argument. Suppose that it is not true. Then there is a nontrivial element x_0 belonging to the orthogonal supplement of $\overline{D(\mathcal{A})}$ such that for all $x \in D(\mathcal{A})$,

$$(x, x_0) = 0. \tag{1.3.14}$$

It follows from (ii) that there is an $x^* \in D(\mathcal{A})$ such that

$$x^* + \mathcal{A}x^* = x_0. \tag{1.3.15}$$

Taking the inner product of (1.3.15) with x^*, we deduce that

$$(x^* + \mathcal{A}x^*, x^*) = 0. \tag{1.3.16}$$

Taking the real part of (1.3.15), we deduce that $x^* = 0$, and by (1.3.16), $x_0 = 0$, a contradiction. Thus the proof is complete. □

Lemma 1.3.13 *Let A be a linear operator with dense domain $D(A)$ in a Hilbert space $\mathcal{H}$. If A is dissipative and $0 \in \rho(A)$ (the resolvent set of A), then A is the infinitesimal generator of a C_0-semigroup of contractions on $\mathcal{H}$.*

Proof See, e.g., Liu and Zheng [155] or Pazy [201]. □

For the abstract initial value problem

$$\begin{cases} \frac{du}{dt} + \mathcal{A}u = K, \\ u(0) = u_0, \end{cases} \tag{1.3.17}$$

where $\mathcal{A}$ is a maximal accretive operator defined in a dense subset $D(\mathcal{A})$ of a Banach space H.

Lemma 1.3.14 *(Pazy [201], Zheng [287]) Suppose that $\mathcal{A}$ is m-accretive in a Banach space H, $K = 0$ and $u_0 \in D(\mathcal{A})$. Then problem (1.3.17) has a unique classical solution u such that*

$$u \in C^1([0, +\infty), H) \cap C([0, +\infty), D(\mathcal{A})).$$

Note that $K = 0$ in (1.3.17) corresponds to the Lumer-Phillips theorem.

Proof By Theorem 1.3.9 and Hille–Yoshida Theorem, $u(t) = S(t)u_0$ is a classical solution in the required class.

To prove the uniqueness, we use a contradiction argument. Suppose that there are two solutions u_1 and u_2. Then

$$u = u_1 - u_2 \in C^1([0, +\infty), H) \cap C([0, +\infty), D(\mathcal{A}))$$

and it satisfies

$$\begin{cases} \frac{du}{dt} + \mathcal{A}u = 0, \\ \\ u(0) = 0, \end{cases} \tag{1.3.18}$$

For any $T > 0$, let

$$\phi(t) = S(t)u(T - t), \quad \forall\, t \in [0, T]. \tag{1.3.19}$$

Then by Theorem 1.3.9, we have

$$\frac{d\phi}{dt} = -S(t)\frac{du(T-t)}{dt} - S(t)\mathcal{A}u(T-t) = S(t)\mathcal{A}u(T-t) - S(t)\mathcal{A}u(T-t) = 0. \tag{1.3.20}$$

Thus,

$$u(T) = \phi(0) = \phi(T) = S(T)u(0) = 0. \tag{1.3.21}$$

and the uniqueness follows. □

Lemma 1.3.15 *(Pazy [201], Zheng [287]) Suppose that $K = K(t)$, and*

$$K(t) \in C^1([0, +\infty), H), \quad u_0 \in D(\mathcal{A}).$$

Then problem (1.3.17) admits a unique global classical solution u such that

$$u \in C^1([0, +\infty), H) \cap C([0, +\infty), D(\mathcal{A})) \tag{1.3.22}$$

which can be described as

$$u(t) = S(t)u_0 + \int_0^t S(t-\tau)K(\tau)d\tau. \tag{1.3.23}$$

Proof Since $S(t)u_0$ satisfies the homogeous equation and nonhomogeneous initial condition, it suffices to verify that $w(t)$ given by

$$w(t) = \int_0^t S(t-\tau)f(\tau)d\tau \tag{1.3.24}$$

belongs to

$$C^1([0, +\infty), H) \cap C([0, +\infty), D(\mathcal{A}))$$

and satisfies the nonhomogeneous equation. Consider the following quotient of difference

$$\begin{aligned} \frac{w(t+h)-w(t)}{h} &= \frac{1}{h}\left(\int_0^{t+h} S(t+h-\tau)f(\tau)d\tau - \int_0^t S(t-\tau)f(\tau)d\tau\right) \\ &= \frac{1}{h}\left(\int_t^{t+h} S(t+h-\tau)f(\tau)d\tau + \frac{1}{h}\int_0^t (S(t+h-\tau)-S(t-\tau))f(\tau)d\tau\right) \\ &= \frac{1}{h}\int_t^{t+h} S(z)f(t+h-z)dz + \frac{1}{h}\int_0^t S(z)(f(t+h-z)-f(t-z))dz, \end{aligned} \tag{1.3.25}$$

when $h \to 0$, the terms in the last line of (1.3.25) have limits

$$S(t)f(0) + \int_0^t S(z)f'(t-z)dz \in C([0, +\infty), H). \tag{1.3.26}$$

It turns out that $w \in C^1([0, +\infty), H)$ and the terms in the third line of (1.3.25) have limits too, which should be

$$S(0)f(t) - \mathcal{A}w(t) = f(t) - \mathcal{A}w(t). \tag{1.3.27}$$

Thus the proof is complete. □

From (1.3.24) we immediately have the following result.

Lemma 1.3.16 *(Pazy [201], Zheng [287]) Suppose that $K = K(t)$, and*

$$K(t) \in C([0, +\infty), D(\mathcal{A})), \quad u_0 \in D(\mathcal{A}).$$

Then problem (1.3.11) admits a unique global classical solution $u(t)$.

Lemma 1.3.17 *(Pazy [201], Zheng [287]) Suppose that $K = K(t)$, and*

$$K(t) \in C([0, +\infty), H), \quad u_0 \in D(\mathcal{A})$$

and for any $T > 0$,

$$K_t \in L^1([0, T], H).$$

Then problem (1.3.17) admits a unique global classical solution $u(t)$.

Proof We first prove that for any $g \in L^1([0, T], H)$, the function w given by the following integral

$$w(t) = \int_0^t S(t - \tau)g(\tau)d\tau \tag{1.3.28}$$

belongs to $C([0, T], H)$. Noting that

$$\begin{aligned} w(t + h) - w(t) &= \int_0^{t+h} S(t + h - \tau)g(\tau)d\tau - \int_0^t S(t - \tau)g(\tau)d\tau \\ &= (S(h) - I)w(t) + \int_t^{t+h} S(t + h - \tau)g(\tau)d\tau, \end{aligned} \tag{1.3.29}$$

we get as $h \to 0$,

$$\| w(t + h) - w(t) \| \le \| (S(h) - I)w(t) \| + \int_t^{t+h} \| g(\tau) \| \to 0 \tag{1.3.30}$$

where we have used the strong continuity of $S(t)$ and the absolute continuity of integral for $\| g \| \in L^1[0, T]$.

Now it can be seen from the last line of (1.3.25) that for almost every $t \in [0, T]$, $\frac{dw}{dt}$ exists and it equals

$$S(t)f(0) + \int_0^t S(z)f'(t-z)dz = S(t)f(0) + \int_0^t S(t-\tau)f'(\tau)d\tau \in C([0,T],H). \tag{1.3.31}$$

Thus, for almost every t,

$$\frac{dw}{dt} = -\mathcal{A}w + f. \tag{1.3.32}$$

Since w and f both belongs to $C([0, T], H)$, it follows from (1.3.32) that for almost every t, $\mathcal{A}w$ belongs to $C([0, T], H)$. Since $\mathcal{A}$ is a closed operator, we conclude that

$$w \in C([0,T], D(\mathcal{A})) \bigcap C^1([0,T],H)$$

and (1.3.32) holds for every t. Thus the proof is complete. □

Lemma 1.3.18 *(Pazy [201], Zheng [287]) When $K = K(u)$ satisfies the global Lipschitz condition, i.e., there is a positive constant L such that for all $u, v \in H$,*

$$\| K(u) - K(v) \|_H \leq \| u - v \|_H . \tag{1.3.33}$$

Furthermore, suppose that $u_0 \in H$. Then problem (1.3.18) admits a global mild solution $u(t)$ such that $u(t)$ belongs to $C([0, +\infty, H)$ and satisfies the following integral equation:

$$u(t) = S(t)u_0 + \int_0^t S(t-\tau)K(u(s))ds. \tag{1.3.34}$$

Moreover, let $u(t)$, $\hat{u}(t)$ be the global mild solutions corresponding to u_0 and $\hat{u}_0$. Then for all $t \geq 0$, the following estimates holds:

$$\| u(t) - \hat{u}(t) \|_H \leq e^{Lt} \| u_0 - \hat{u}_0 \|_H . \tag{1.3.35}$$

Proof We use the contraction mapping theorem to prove the present theorem. Two key steps for using the contraction mapping theorem are to figure out a closed set of the considered Banach space, and an auxiliary problem so that the nonlinear operator defined by the auxiliary problem maps from this closed set into itself, and turns out to be a contraction. In the following we proceed along this line.

Let

$$\phi(v) = S(t)u_0 + \int_0^t S(t-\tau)K(v(t))d\tau \tag{1.3.36}$$

and

$$X = \{v \in C([0, +\infty), H \,|\, \sup_{t\geq 0} \| v(t) \| e^{-kt} < +\infty\} \tag{1.3.37}$$

where k is a positive constant such that $k > L$. In X, we introduce the following norm:

$$\| u \|_X = \sup_{t\geq 0} e^{-kt} \| u(t) \| . \tag{1.3.38}$$

Clearly, X is a Banach space. We now show that the nonlinear operator ϕ defined by (1.3.36) maps X into itself, and the mapping is a contraction. In deed, for $v \in X$, by (1.3.33), we have

$$\begin{aligned} \| \phi(v) \| &\leq \| S(t)u_0 \| + \int_0^t \| S(t-\tau) \| \| K(v(t)) \| d\tau \\ &\leq \| u_0 \| + C_0 t + L \sup_{t\geq 0} e^{-kt} \| v(t) \| \cdot \int_0^t e^{k\tau} d\tau \\ &\leq \| u_0 \| + C_0 t + \frac{L}{k} e^{kt} \| v(t) \|_X . \end{aligned} \tag{1.3.39}$$

where $C_0 = K(0)$. Thus,

$$\| \phi(v) \|_X \leq \sup_{t\geq 0} \left((\| u_0 \| + C_0 t) e^{-kt} \right) + \frac{L}{k} \| v(t) \|_X < +\infty. \tag{1.3.40}$$

i.e., $\phi(v) \in X$.

For $v_1, v_2 \in X$, we have

$$\begin{aligned} \| \phi(v_1) - \phi(v_2) \|_X &= \sup_{t\geq 0} e^{-kt} \| \int_0^t S(t-\tau)(K(v_1(\tau)) - K(v_2(\tau)))d\tau \| \\ &\leq \sup_{t\geq 0} e^{-kt} L \int_0^t \| v_1 - v_2 \| d\tau \leq \frac{L}{k} \| v_1 - v_2 \|_X . \end{aligned} \tag{1.3.41}$$

Therefore, by the contraction mapping theorem, (1.3.34) has a unique solution in X. To show that the uniqueness also holds in $C([0, +\infty), H)$, let $u_1, u_2 \in C([0, +\infty), H)$ be two solutions of (1.3.34) and let $u = u_1 - u_2$. Then

$$u(t) = \int_0^t S(t-\tau)(K(u_1) - K(u_2))d\tau, \tag{1.3.42}$$

$$\| u(t) \| \leq L \int_0^t \| u(\tau) \| d\tau. \tag{1.3.43}$$

By the Gronwall inequality, we immediately conclude that $u(t) = 0$, i.e., the uniqueness in $C([0, +\infty), H)$ follows.

It remains to prove the stability result (1.3.35). In the same way as above, we have

$$u(t) - \hat{u}(t) = S(t)(u_0 - \hat{u}_0) + \int_0^t S(t-\tau)(K(u) - K(\hat{u}))d\tau. \quad (1.3.44)$$

Therefore,

$$\| u(t) - \hat{u}(t) \| \leq \| u_0 - \hat{u}_0 \| + L \int_0^t \| u(\tau) - \hat{u} \| d\tau. \quad (1.3.45)$$

By the Gronwall inequality, (1.3.35) follows. Thus the proof is complete. □

For the abstract initial value problem

$$\begin{cases} \dfrac{du}{dt} = \mathcal{A}u, \\ u(0) = u_0, \end{cases}$$

where $\mathcal{A}$ was defined as in Theorem 1.3.11, we define a continuous non-decreasing function

$$M(\eta) = \max_{t \in [-\eta, \eta]} \| R(it, \mathcal{A}), \|, \quad \eta > 0 \quad (1.3.46)$$

where $R(\lambda, \mathcal{A}) = (\lambda I - \mathcal{A})^{-1}$, I is a unit operator on the Banach space H. Then the associated function

$$M_{\log}(\eta) := M(\eta)(\log(1 + M(\eta)) + \log(1 + \eta)), \quad \eta > 0. \quad (1.3.47)$$

Let $M_{\log}^{-1}$ be the inverse of M_{log} defined on $[M_{\log}(0), +\infty)$. Then, we have the following lemmas.

Lemma 1.3.19 *(Batty and Duyckaerts [13]) Let $S(t), t \geq 0$ be a bounded C_0-semigroup on a Banach space X with generator $\mathcal{A}$, such that $i\mathbb{R} \subset \rho(\mathcal{A})$. Let the functions M and $M_{\log}$ be defined by (1.3.46) and (1.3.47). Then there exist constants $C, B > 0$ such that for all $t \geq B$,*

$$\| S(t)\mathcal{A}^{-1} \| \leq \frac{C}{M_{\log}^{-1}(t/C)}. \quad (1.3.48)$$

If $\alpha > 0, M(\eta) \leq C(1 + \eta^\alpha), \ \eta \geq 0$, then for all $t \geq B$,

$$\| S(t)\mathcal{A}^{-1} \| \leq C \left(\frac{\log t}{t} \right)^{\frac{1}{\alpha}}. \quad (1.3.49)$$

Proof Let $t \geq 0$ and $R > 0$, and let γ be the contour consisting of the right-hand of the circle $|z| = R$ and any path γ' in $\{z \in \rho(-\mathcal{A}) : Rez < 0\}$ from $i\mathbb{R}$ to $i\mathbb{R}$. By Cauchy's Theorem,

$$e^{-t\mathcal{A}}\mathcal{A}^{-1} = \frac{1}{2\pi i}\int_{\gamma}\left(1+\frac{z^2}{R^2}\right)(z+\mathcal{A})^{-1}e^{-t\mathcal{A}}\frac{dz}{z}.$$

We first show that

$$\| e^{-t\mathcal{A}}\mathcal{A}^{-1} \| \leq \frac{2\tilde{C}}{R} + \frac{1}{2\pi} \| \int_{\gamma'}\left(1+\frac{z^2}{R^2}\right)(z+\mathcal{A})^{-1}e^{tz}\frac{dz}{z} \| . \quad (1.3.50)$$

If $z = Re^{i\theta}$, $-\pi/2 < \theta < \pi/2$, we have

$$\| (z+\mathcal{A})^{-1}e^{-t\mathcal{A}} \| = \| e^{tz}\int_t^{+\infty} e^{-(z+\mathcal{A})s}ds \| \leq \frac{\tilde{C}}{R\cos\theta}.$$

Noticing that $|1 + z^2/R^2| = 2|\cos\theta|$, we get the bound

$$\| \int_{|z|=R}\left(1+\frac{z^2}{R^2}\right)(z+\mathcal{A})^{-1}e^{-t\mathcal{A}}\frac{dz}{z} \| \leq \frac{2\pi\tilde{C}}{R}. \quad (1.3.51)$$

Next, define the analytic function of z

$$h_t(z) = \int_0^t e^{(t-sz)z}e^{-s\mathcal{A}}ds = (z+\mathcal{A})e^{tz} - (z+\mathcal{A})^{-1}e^{-t\mathcal{A}}. \quad (1.3.52)$$

The Cauchy's Theorem and similar estimate as for (1.3.51) yield

$$\| \int_{\gamma'}\left(1+\frac{z^2}{R^2}\right)h_t(z)\frac{dz}{z} \| = \| \int_{|z|=R}\left(1+\frac{z^2}{R^2}\right)h_t(z)\frac{dz}{z} \| \leq \frac{2\pi\tilde{C}}{R}, \quad (1.3.53)$$

which, together with (1.3.52) and (1.3.53), yields the estimate (1.3.50).

By means of standard Neumann series, we may take γ' to be the union of γ_0, γ_+ and γ_-, where

$$\gamma_0(\tau) = -\frac{1}{M(|\tau|)} + i\tau \quad (-R \leq \tau \leq R), \quad \gamma_\pm(s) = s \pm iR \quad (-(2M(R))^{-1} \leq s < 0).$$

Although γ_0 may not be piecewise smooth, it can be approximated by smooth paths and (1.3.50) remains valid. Moreover,

$$\| (\gamma_0(\tau) + \mathcal{A}^{-1}) \| \leq 2M(|\tau|), \quad \| (\gamma_\pm(s) + \mathcal{A}^{-1}) \| \leq 2M(R).$$

Now assume that $R > 1$. On $\gamma_\pm$, $|1 + z^2/R^2| \leq C/R$, so we can estimate the norms of the integral in (1.3.50) over $\gamma_\pm$ by

$$C \int_0^{(2M(R))^{-1}} \frac{1}{R} M(R) \frac{e^{-ts}}{R} ds \leq \frac{CM(R)}{R^2 t},$$

where $C > 0$ is a constant depending only on M. On γ_0, $1+z^2/R^2$ and $1/z$ are bounded independently of R, and $e^{tz} \leq e^{-t/2M(R)}$. The length of γ_0 is at most $2(M(0) + R)$. Hence, we can estimate the norm of the integral over γ_0 by

$$CM(R)(1 + R)e^{-t/2M(R)},$$

where again $C > 0$ depends only on M.

Thus we have the estimate

$$\| e^{-t\mathcal{A}} \mathcal{A}^{-1} \| \leq C\Big(\frac{1}{R} + \frac{M(R)}{R^2 t} + \frac{(1 + M(R))^2(1 + R)^2}{R} e^{-t/2M(R)}\Big).$$

Here, C depends on $\tilde{C}$ as well as M. Given $t > 4M_{\log}(1)$, choose $R = M_{\log}^{-1}(t/4)$. Then

$$\frac{M(R)}{R^2 t} = \frac{1}{4R^2 \log((1 + M(R))(1 + R))} \leq \frac{C}{R},$$
$$(1 + M(R))^2(1 + R)^2 e^{-t/2M(R)} = 1.$$

Hence, the required result is proved. □

Lemma 1.3.20 (Borichev and Tomilov [19]) *Given $Q > 0$ and $\varepsilon > 0$, there exist an integer $k > Q$ and a complex measure μ with compact support in $\mathbb{C}\backslash\Omega$ such that for some $B = B(\alpha, \beta)$, we have*

$$\Big|\int_{\mathbb{C}/\Omega} \frac{d\mu(\zeta)}{z - \zeta}\Big| \leq B(1 + |Imz|)^\beta \cdot \chi_{\{\xi:|\xi|>Q\}}(z) + \varepsilon, \quad z \in \Omega, \tag{1.3.54}$$

$$\Big|\int_{\mathbb{C}/\Omega} e^{t\zeta} d\mu(\zeta)\Big| \leq B\chi_{\{r:|r|>Q\}}(t) + \varepsilon, \quad t \geq 0, \tag{1.3.55}$$

$$1/B \leq \Big|\int_{\mathbb{C}\backslash\Omega} e^{t\zeta} d\mu(\zeta)\Big| \leq B, \tag{1.3.56}$$

$$\Big|\int_{\mathbb{C}\backslash\Omega} e^{t\zeta} \frac{d\mu(\zeta)}{\zeta}\Big|^\alpha \leq B\frac{\log t}{t} \cdot \chi_{(Q,2k)}(t) + \frac{\varepsilon}{t+1}, \quad t \geq 0, \tag{1.3.57}$$

and

$$\Big|\int_{\mathbb{C}\setminus\Omega} e^{k\zeta}\frac{d\mu(\zeta)}{\zeta}\Big|^{\alpha} \geq \frac{\log k}{Bk}. \tag{1.3.58}$$

Proof Choose $H > Q$ large enough, and for an integer k, $k \geq 2$, such that

$$H^{\alpha} \leq k \leq H^{3\alpha/2},$$

define

$$A = 2k\log k, \quad \tau = A^{k-1}/\sqrt{k}, \quad q = e^{2\pi i/k}, \quad w = iH - 1.$$

We define also a finite measure

$$\mu = \tau \sum_{1\leq s\leq k} q^s(1 + q^s/(Aw))\delta_{w+q^s/A},$$

where δ_x is the unit mass at x.

Observe that

$$\sum_{1\leq s\leq k}\frac{q^s}{x-q^s} = \frac{k}{x^k-1}, \quad \sum_{1\leq s\leq k}\frac{q^{2s}}{x-q^s} = \frac{kx}{x^k-1}. \tag{1.3.59}$$

Indeed, to prove the fist identity we use that

$$\sum_{1\leq s\leq k}\frac{q^s}{x-q^s} = \frac{P(x)}{x^k-1},$$

for some polynomial P with $\deg P < k$ such that $P(x) = P(qx)$ so that $P(x) = \text{const.}$, and then $P(x) = P(0) = k$. The second equality can be proved in a similar way by using that if

$$\sum_{1\leq s\leq k}\frac{q^{2s}}{x-q^s} = \frac{P(x)}{x^k-1},$$

then $P(qx) = qP(x)$ which yields $P(x) = kx$.

Now using (1.3.59), we get

$$\begin{aligned}\mathbb{C}_{\mu(z)} &= \int_{\mathbb{C}/\Omega}\frac{d\mu(\zeta)}{z-\zeta} = \tau\sum_{1\leq s\leq k}\frac{q^s(1+q^s/(Aw))}{(z-w)-q^s/A}\\ &= \tau\sum_{1\leq s\leq k}\Big[\frac{Aq^s}{A(z-w)-q^s} + \frac{q^{2s}/w}{A(z-w)-q^s}\Big]\end{aligned}$$

$$= \tau \frac{Ak}{A^k(z-w)^k - 1} + \frac{\tau}{w} \frac{kA(z-w)}{A^k(z-w)^k - 1}$$
$$= \frac{z}{w} \frac{\tau Ak}{A^k(z-w)^k - 1}.$$

If $z \in \Omega, H > 1$, then

$$|z - w| \geq 1 - \frac{1}{H^\alpha}.$$

Now if $\frac{H}{2} \leq Imz \leq |z| \leq 2H, z \in \Omega$, then we use that $A|z-w| > 2$, $|A^k(z-w)^k - 1| > A^k e^{-k/H^\alpha}/2$, to obtain that

$$\left| \frac{z}{w} \frac{\tau Ak}{A^k(z-w)^k - 1} \right| \leq c\tau k A^{1-k} e^{k/H^\alpha} = c\sqrt{k} e^{k/H^\alpha}.$$

From now on, we assume that k satisfies the condition

$$\sqrt{k} e^{k/H^\alpha} \leq H^\beta. \tag{1.3.60}$$

Then

$$|\mathcal{C}_{\mu(z)}| \leq cH^\beta, \quad \frac{H}{2} \leq Imz \leq |z| \leq 2H, \quad z \in \Omega. \tag{1.3.61}$$

If $z \in \Omega$ and $|Imz - H| + ||z| - H| > \frac{H}{2}$, then $|z - w| > c\max(|z|, H)$, and under condition (1.3.60), we have for large H,

$$|\mathcal{C}_{\mu(z)}| \leq \frac{c_1|z|\sqrt{k}}{H(c\max(|z|, H))^k} \leq \frac{c_1|z|\sqrt{k}}{H(c\max(|z|, H))^{\alpha+1}} \leq \varepsilon.$$

This, together with (1.3.61), proves (1.3.54).

Next, define

$$\mathcal{L}_{\mu(t)} = \int_{\mathbb{C}\setminus\Omega} e^{t\zeta} d\mu(\zeta) = \tau \sum_{1\leq s\leq k} q^s(1 + q^s/(Aw))e^{t(w+q^s/A)},$$

and

$$|\mathcal{L}_{\mu(t)}| = \tau e^{-t} \left| \sum_{1\leq s\leq k} q^s(1 + q^s/(Aw))e^{t(q^s t/A)} \right|. \tag{1.3.62}$$

Furthermore, we have

$$\begin{aligned}
&\sum_{1\le s\le k} q^s(1+q^s/(Aw))e^{t(q^s t/A)}\\
&=\sum_{1\le s\le k}\sum_{n\ge 0}(q^s+q^{2s}/(Aw))(q^s t/A)^n\frac{1}{n!}\\
&=k\sum_{m\ge 1}\left[\frac{t^{km-1}}{A^{km-1}}\cdot\frac{1}{(km-1)!}+\frac{t^{km-2}}{A^{km-2}}\cdot\frac{1}{(km-2)!}\cdot\frac{1}{Aw}\right]\\
&=\frac{kt^{k-1}}{A^{k-1}(k-1)!}\sum_{m\ge 1}\left(\frac{t^k}{A^k}\right)^{m-1}\frac{(k-1)!}{(km-1)!}\left(1+\frac{km-1}{tw}\right),
\end{aligned}$$

and

$$|\mathcal{L}_{\mu(t)}|=\frac{k^{3/2}t^{k-1}e^{-t}}{k!}\sum_{m\ge 1}\left(\frac{t^k}{A^k}\right)^{m-1}\frac{(k-1)!}{(km-1)!}\left(1+\frac{km-1}{tw}\right).$$

Thus, for some constants c, c_1, c_2, c_3, we have

$$\begin{aligned}
|\mathcal{L}_{\mu(t)}| &\le c\frac{k^{3/2}t^{k-1}e^{-t}}{k!}\sum_{m\ge 1}\left(\frac{t^k}{HA}\right)^{k(m-1)}\frac{(k-1)!}{(km-1)!}\frac{km}{H}\\
&\le c_1 e^{-t}k^{5/2}t^{k-2}/(k!H),\quad 0\le t\le k/H,
\end{aligned}$$

and

$$c_2 e^{-t}k^{3/2}t^{k-1}/(k!)\le|\mathcal{L}_{\mu(t)}|\le c_3 e^{-t}k^{3/2}t^{k-1}/(k!),\quad k/H\le t\le A.$$

If $t\ge A$, then by (1.3.62),

$$|\mathcal{L}_{\mu(t)}|\le 2\tau e^{-t}ke^{t/A}=2\sqrt{k}A^{k-1}e^{-t(1-1/A)}.$$

The function $t\mapsto e^{-t}t^{k-2}$ attains its maximum on $[0,k/H]$ at $t=k/H$; the function $t\mapsto e^{-t}t^{k-1}$ attains its maximum on $[k/H,k]$ at $t=k-1$; the function $t\mapsto e^{-t(1-1/A)}$ attains its maximum on $[A,+\infty)$ at $t=A$. Using that $A=2k\log k$, by Stirling's formula, we obtain that

$$\begin{aligned}
&0<c_1\le|\mathcal{L}_{\mu(k)}|\le c_2|\mathcal{L}_{\mu(k-1)}|\le c_3,\quad \max_{[k/H,k/2]}|\mathcal{L}_\mu|=\mathrm{o}(1),\quad H\to+\infty,\\
&e^{-k/H}k^{5/2}(k/H)^{k-2}/(k!H)\to 0,\quad \sqrt{k}A^{k-1}e^{-A}\to 0,\quad H\to+\infty.
\end{aligned}$$

Hence,

$$0 < c_1 \le |\mathcal{L}_{\mu(k)}| \le c_2 \max_{\mathbb{R}^+} |\mathcal{L}_\mu| \le c_3, \qquad \max_{[0,k/2]\bigcup[A,+\infty)} |\mathcal{L}_\mu| = \mathrm{o}(1), \quad H \to +\infty,$$

and (1.3.55) and (1.3.56) follow for large H.

Finally, define

$$\mathcal{N}_{\mu(t)} = \int_{\mathbb{C}\setminus\Omega} e^{t\zeta} \frac{d\mu(\zeta)}{\zeta} = \frac{\tau e^{-t+iHt}}{w} \sum_{1\le s\le k} q^s e^{q^s t/A}.$$

Here we use the formula

$$\begin{aligned}\sum_{1\le s\le k} q^s e^{q^s t/A} &= \sum_{1\le s\le k} \sum_{n\ge 0} q^s (q^s t/A)^n \frac{1}{n1} \\ &= \frac{kt^{k-1}}{A^{k-1}(k-1)!} \sum_{m\ge 1} \left(\frac{t^k}{A^k}\right)^{m-1} \frac{(k-1)1}{(km-1)!} \\ &= (1+\mathrm{o}(1)) \frac{kt^{k-1}}{A^{k-1}(k-1)!}, \quad 0 \le t \le A, \quad H \to +\infty.\end{aligned}$$

Therefore, by Stirling's formula, we have for large H:

$$|\mathcal{N}_{\mu(k)}| \ge c_1 \frac{\tau e^{-k}}{H} \cdot \frac{k^k}{A^{k-1}(k-1)1} \ge \frac{c_2}{H}. \tag{1.3.63}$$

Moreover,

$$t^{1/\alpha} |\mathcal{N}_{\mu(t)}| \le \varepsilon + c_1 (k^{1/\alpha}/H) \cdot \chi_{(k/2,2k)}(t), \quad 0 \le t \le A, \tag{1.3.64}$$

and

$$t^{1/\alpha} |\mathcal{N}_{\mu(t)}| \le \frac{c\tau e^{-t}}{H} \cdot kt^{1/\alpha} e^{t/A} \le c_1 \frac{A^{k-1+1/\alpha}\sqrt{k} e^{-A}}{H} \le \varepsilon, \quad t \ge A. \tag{1.3.65}$$

Now we fix $0 < \psi < \beta - \frac{\alpha}{2}$ and $k = \psi H^\alpha \log H$ in such a way that $k \in \mathbb{N}$. Then (1.3.60) is satisfied for large H, (1.3.63) implies (1.3.58), and (1.3.64)–(1.3.65) imply (1.3.57). □

Lemma 1.3.21 (Borichev and Tomilov [19]) *Given $\alpha > 0$, there exist a Banach space X_α and a bounded C_0-semigroup $S(t)$ on X_α with generator $\mathcal{A}$ such that*

(i)

$$\| R(is, \mathcal{A}) \| = O(|s|^\alpha), \quad |s| \to +\infty$$

(ii)

$$\lim_{t\to+\infty} \sup\left(\frac{t}{\log t}\right)^{\frac{1}{\alpha}} \| S(t)\mathcal{A}^{-1} \| > 0.$$

Proof Let $T(t)_{t\geq 0}$ be the left shift semigroup on BUC$(\mathbb{R}^+, H)$, where BUC$(\mathbb{R}^+, H)$ is the space of bounded uniformly continuous H-valued functions. Let $\Omega := \{\lambda \in \mathbb{C} : \mathrm{Re}\lambda > -1/(1+|\mathrm{Im}\lambda|)^\alpha\}$, and let $\Omega_0 := \Omega \bigcap \{\lambda \in \mathbb{C} : |\mathrm{Re}\lambda| < 1$. Furthermore, let H_α be the space of functions $f \in$ BUC$(\mathbb{R}^+)$ such that the Laplace transform $\hat{f}$ extends to an analytic function in Ω_0, and

$$|\hat{f}(\lambda)|(1+|\mathrm{Im}\lambda|)^{-\alpha} \to 0, \quad \lambda \to \infty,\ \lambda \in \Omega_0.$$

Then H_α equipped with the norm

$$\| f \|_{H_\alpha} =: \| f \|_\infty + \| f \|_\alpha = \| f \|_\infty + \sup_{\lambda\in\Omega_0} |\hat{f}(\lambda)|(1+|\mathrm{Im}\lambda|)^{-\alpha}$$

is a Banach space. Moreover, $T(t)H_\alpha \subset H_\alpha, t \geq 0$, and the restriction $S(t)$ of $T(t)$ to H_α is also a C_0-semigroup. To prove this assertion, it suffices to observe that

$$\begin{aligned}\hat{f}(\lambda) - \widehat{S(t)f}(\lambda) &= \int_0^{+\infty} e^{-\lambda s} f(s)ds - \int_0^{+\infty} e^{-\lambda s} f(t+s)ds\\ &= (1-e^{\lambda t})\hat{f}(\lambda) + e^{\lambda t}\int_0^t e^{-\lambda s} f(s)ds, \quad \mathrm{Re}\lambda > 0,\end{aligned}$$

and the same equality holds on Ω_0. By the definition of H_α,

$$\| S(t)f - f \|_\alpha \to 0, \quad t \to 0^+,$$

and then

$$\| S(t)f - f \|_{H_\alpha} \to 0, \quad t \to 0^+.$$

Let $\mathcal{A}$ stand for the generator of semigroup $S(t)$. Next we prove that for every $f \in H_\alpha$, the local resolvent $R(\lambda, \mathcal{A})f$ satisfies the estimate

$$\| R(\lambda, \mathcal{A})f \|_{H_\alpha} \leq C(1+|\mathrm{Im}\lambda|)^\alpha \| f \|_{H_\alpha}, \quad 0 < \mathrm{Re}\lambda < 1. \tag{1.3.66}$$

We will estimate the quantities $\| R(\lambda, \mathcal{A})f \|_\infty$ and $\| R(\lambda, \mathcal{A})f \|_\alpha$ separately.

Observe first that for every $t \in \mathbb{R}^+$ and every $\lambda \in \mathbb{C}_+$, one has

$$(R(\lambda, \mathcal{A})f)(t) = \int_0^{+\infty} e^{-\lambda s} f(t+s)ds = e^{\lambda t}\hat{f}(\lambda) - \int_0^t e^{\lambda(t-s)} f(s)ds.$$

It follows that for every fixed $t \in \mathbb{R}$, the function $\lambda \mapsto (R(\lambda, \mathcal{A})f)(t)$ extends to an analytic function on Ω_0 and moreover

$$|(R(\lambda, \mathcal{A})f)(t)| \le \frac{\| f \|_\infty}{|\mathrm{Re}\lambda|}, \quad \text{if } \mathrm{Re}\lambda > 0,$$
$$|(R(\lambda, \mathcal{A})f)(t)| \le \frac{\| f \|_\infty}{|\mathrm{Re}\lambda|} + |\hat{f}(\lambda)|, \quad \text{if } \mathrm{Re}\lambda < 0.$$

Applying Levinson's log-log theorem, we conclude that

$$\| R(\lambda, \mathcal{A}) \|_\infty \le C(1 + |\mathrm{Im}\lambda|)^\alpha (\| f \|_\infty + \| f \|_\alpha), \quad 0 < \mathrm{Re}\lambda < 1. \tag{1.3.67}$$

Fix $\lambda \in (0, 1)$. To estimate $\| R(\lambda, \mathcal{A})f \|_\alpha$, note that

$$\begin{aligned}
(\widehat{R(\lambda, \mathcal{A})}f)(\mu) &= \int_0^{+\infty} e^{-\mu t} \int_0^{+\infty} e^{-\lambda s} f(t+s)dsdt \\
&= \int_0^{+\infty} e^{(\lambda-\mu)t} \int_t^{+\infty} e^{-\lambda s} f(s)dsdt \\
&= -\frac{1}{\lambda - \mu} \int_0^{+\infty} e^{-\lambda s} f(s)ds + \frac{1}{\lambda - \mu} \int_0^{+\infty} e^{-\mu t} f(t)dt \\
&= -\frac{\hat{f}(\lambda) - \hat{f}(\mu)}{\lambda - \mu}, \quad \mathrm{Re}\,\mu > 1.
\end{aligned}$$

Therefore, $\widehat{R(\lambda, \mathcal{A})}f$ extends analytically to Ω_0, and

$$\begin{aligned}
(\widehat{R(\lambda, \mathcal{A})}f)(\mu) &= -\frac{\hat{f}(\lambda) - \hat{f}(\mu)}{\lambda - \mu}, \quad \lambda \ne \mu, \ \mu \in \Omega_0, \\
(\widehat{R(\lambda, \mathcal{A})}f)(\mu) &= -\hat{f}'(\mu), \qquad \lambda = \mu.
\end{aligned}$$

Now, if $|\lambda - \mu| \ge 1$, $\mu \in \Omega_0$, then

$$|\widehat{R(\lambda, \mathcal{A})}f(\mu)| \le |\hat{f}(\lambda)| + |\hat{f}(\mu)| \le c(1 + |\mathrm{Im}\mu|)^\alpha (1 + |\mathrm{Im}\lambda|)^\alpha \| f \|_\alpha .$$

Furthermore, if

$$\frac{1}{2(1 + |\mathrm{Im}\lambda|)^\alpha} \le |\lambda - \mu| \le 1, \quad \mu \in \Omega_0,$$

then we have

$$\begin{aligned}
|\widehat{R(\lambda, \mathcal{A})}f(\mu)| &\le 2(1 + |\mathrm{Im}\lambda|)^\alpha |\hat{f}(\lambda)| + |\hat{f}(\mu)| \\
&\le c(1 + |\mathrm{Im}\mu|)^\alpha (1 + |\mathrm{Im}\lambda|)^\alpha \| f \|_\alpha .
\end{aligned}$$

Finally, if

$$|\lambda - \mu| \le \frac{1}{2(1 + |\mathrm{Im}\lambda|)^\alpha},$$

then, applying Cauchy's formula on the circle

$$C_\lambda := \left\{ z \in \mathbb{C} : |z - \lambda| = \frac{2}{3}(1 + |\text{Im}\lambda|)^{-\alpha} \right\},$$

we obtain that

$$|\widehat{R(\lambda, \mathcal{A})f}(\mu)| \leq c(1 + |\text{Im}\mu|)^\alpha (1 + |\text{Im}\lambda|)^\alpha \parallel f \parallel_\alpha .$$

Thus,

$$\parallel R(\lambda, \mathcal{A})f \parallel_\alpha \leq C(1 + |\lambda|)^\alpha \parallel f \parallel_\alpha, \quad 0 < \text{Re}\lambda < 1. \tag{1.3.68}$$

The estimates (1.3.67) and (1.3.68) together give us (1.3.66). Since

$$\parallel R(\lambda, \mathcal{A})f \parallel_{H_\alpha} \geq \frac{1}{\text{dist}(\lambda, \sigma(\mathcal{A}))},$$

the estimate (1.3.66) implies that $i\mathbb{R} \subset \mathbb{C} \backslash \sigma(\mathscr{A})$, and that

$$\parallel R(\lambda, \mathcal{A})f \parallel_{H_\alpha} \leq C(1 + |\text{Im}\lambda|)^\alpha, \quad \lambda \in i\mathbb{R}. \tag{1.3.69}$$

Since $\sigma(\mathcal{A}) \cap i\mathbb{R} = \varnothing$, by Theorem 1.3.11, we obtain, for every $f \in H_\alpha$,

$$\mathcal{A}^{-1}f = \lim_{t \to +\infty} \left(\mathcal{A}^{-1}f - S(t)\mathcal{A}^{-1}f \right) = -\int_0^{+\infty} S(t)fdt.$$

By Lemma 1.3.20, there exist $f_n \in H_\alpha, f_n = \mathcal{L}_{\mu_n}$, and $k_n \to +\infty$ as $n \to +\infty$, such that

$$\parallel f_n \parallel_{H_\alpha} \leq 1, \quad |\mathcal{N}\mu_n(k_n)| \geq C(\alpha) \left(\frac{\log k_n}{k_n} \right)^{1/\alpha}, \quad n \geq 1.$$

Therefore,

$$\begin{aligned} \parallel S(k_n)\mathcal{A}^{-1} \parallel_{H_\alpha} &\geq \parallel S(k_n)\mathcal{A}^{-1}f_n \parallel_{H_\alpha} \geq \left| \int_{k_n}^{+\infty} f_n(r)dr \right| \\ &\geq \left| \hat{f}_n(0) - \int_0^{k_n} f_n(s)ds \right| = |\mathcal{N}_{\mu_n}(k_n)| \geq C(\alpha) \left(\frac{\log k_n}{k_n} \right)^{1/\alpha}. \end{aligned}$$

□

Lemma 1.3.22 *A semigroup of contractions $\{e^{t\mathcal{A}}\}_{t \geq 0}$ in a Hilbert space with norm $\parallel \cdot \parallel$ is exponentially stable if and only if*

(i) the resolvent set $\rho(\mathcal{A})$ of $\mathcal{A}$ contains the imaginary axis and

(ii)

$$\lim_{\lambda \to +\infty} \sup \parallel (i\lambda Id - \mathcal{A})^{-1} \parallel < +\infty$$

hold.

Proof The only if part is obvious. We shall prove the if part. Let $\omega = \tau + i\lambda,\ \tau \in [-1/2M, 0]$, $M = \sup\{\| (i\lambda I - \mathcal{A})^{-1} \|\}$, $M_0 = \sup\{\| (\omega I - \mathcal{A})^{-1} \|\}$. Then, we have

$$\begin{aligned}\| [I + \tau(i\lambda - \mathcal{A})^{-1}]x \| &\geq \| x \| - \frac{1}{2M} \| (i\lambda - \mathcal{A})^{-1}x \| \\ &\geq \| x \| - \frac{1}{2M} \| x \| = \frac{\| x \|}{2},\end{aligned}$$

then we have $\| [I + \tau(i\lambda - \mathcal{A})^{-1}]x \| \leq 2$. Thus, for $\tau \in (-1/2M, 0]$, $-\infty < \lambda < +\infty$, we have $\omega \in \rho(\mathcal{A})$ and

$$\begin{aligned}\| (\omega I - \mathcal{A}) \| &= \| (i\lambda I - \mathcal{A})^{-1}[I + \tau(i\lambda I - \mathcal{A})^{-1}]^{-1} \| \\ &\leq \| (i\lambda I - \mathcal{A})^{-1} \| \| [I + \tau(i\lambda I - \mathcal{A})^{-1}]^{-1} \| \leq 2M.\end{aligned}$$

Hence, we have

$$\{\omega : -1 \leq 2M_0 \leq \mathrm{Re}\omega \leq 1/2M_0\} \subset \rho(\mathcal{A}) \tag{1.3.70}$$

and

$$\sup\{\| (\omega I - \mathcal{A})^{-1} \| : -1/2M_0 \leq \mathrm{Re}\omega \leq 1/2M_0\} \leq 2M_0. \tag{1.3.71}$$

On the other hand, from $\| e^{t\mathcal{A}} \| \leq M$, $\{\omega : \mathrm{Re}\omega \subset \rho(\mathcal{A})\}$ and $\sup\{\| (\omega I - \mathcal{A})^{-1} \| : \mathrm{Re}\omega \geq 1/2M_0\} \leq 2M_0 M$, we have

$$\sup\{\| (\omega I - \mathcal{A})^{-1} \| : \mathrm{Re}\omega \geq 1/2M_0\} \leq 2M_0(1 + M),$$

then, set $\mathcal{A}_1 = \mathcal{A} + \frac{1}{M_0}I$, we have there exists a positive constant number M such that $\| e^{t\mathcal{A}_1} \| \leq M$, thus $\| e^{t\mathcal{A}} \| \leq Me^{-\frac{1}{M_0}t}$. The theorem is proved.

Lemma 1.3.23 *If a bounded C_0-semigroup $e^{t\mathcal{A}}$ on a Hilbert space $\mathcal{H}$ satisfies*

$$i\mathbb{R} \subset \rho(\mathcal{A}), \quad \sup_{|\beta| \geq 1} \frac{1}{\beta^j} \| (i\beta I - \mathcal{A})^{-1} \|_{\mathcal{H}} < +\infty, \tag{1.3.72}$$

for some $j > 0$, then for any positive integer m there exists a constant $C_m > 0$ such that

$$\| e^{t\mathcal{A}} z_0 \|_{\mathcal{H}} \leq C_m \left(\frac{\ln t}{t}\right)^{\frac{m}{j}} (\ln t) \| z_0 \|_{D(\mathcal{A}^m)} \tag{1.3.73}$$

for all $z_0 \in D(\mathcal{A}^m)$.

For more detailed proof, we thus refer to Huang [118].

Chapter 2
Exponential Stability for Nonlinear Thermoelastic Equations with Second Sound

2.1 Introduction

This chapter is concerned with the global existence and asymptotic behavior of solutions to the equations of one-dimensional nonlinear thermoelasticity with thermal memory and second sound. We adopt the results in this chapter from [223].

The reference configuration under consideration is the unit interval $\Omega = (0, 1)$. The equations under consideration read as follows

$$\begin{cases} u_{tt} - S(u_x, \theta)_x = 0, & (2.1.1) \\ \theta_t + \gamma q_x + k_1 * q_x + \beta u_{xt} = 0, & (2.1.2) \\ q_t + q + k\theta_x = 0 & (2.1.3) \end{cases}$$

subject to the boundary conditions

$$u(0, t) = u(1, t) = q(0, t) = q(1, t) = 0 \tag{2.1.4}$$

and initial conditions

$$u(x, 0) = u_0(x),\ u_t(x, 0) = u_1(x),\ \theta(x, 0) = \theta_0(x),\ q(x, 0) = q_0(x). \tag{2.1.5}$$

Here by $u = u(x, t)$, $\theta = \theta(x, t)$ and $q = q(x, t)$, we denote the displacement, absolute temperature and heat flux respectively, $S(u_x, \theta)$ is the Piola-Kirchhoff stress tensor, $k_1 = k_1(t)$ is the relaxation kernel. The sign $*$ denotes the convolution product, i.e., $k_1 * y(\cdot, t) = \int_0^t k_1(t - \tau)y(\cdot, \tau)d\tau$. Finally, $\alpha,\ \beta, \gamma$ are positive constants.

When the problem (2.1.1)–(2.1.3) has no the relaxation kernel k_1 (i.e., $k_1 = 0$), Messaoud and Said-Houari [169] considered a one-dimensional homogeneous body occupying, in its reference configuration, an interval I, the laws of balance of momentum, balance of energy, and growth of entropy with the forms

Y. Qin and Z. Ma, *Global Well-posedness and Asymptotic Behavior of the Solutions to Non-classical Thermo(visco)elastic Models*,
DOI 10.1007/978-981-10-1714-8_2

$$\begin{cases} \rho u_{tt} = \sigma_x + b, \\ e_t + q_x = \sigma \varepsilon_t + r, \\ \eta_t \geq \frac{r}{\theta} - \left(\frac{q}{\theta}\right)_x, \end{cases}$$

where the displacement u, the strain $\varepsilon = u_x$, the stress σ, the 'absolute' temperature θ, the heat flux q, the internal energy e, the body force b, and the external heat supply r are all functions of (x, t) $(t \geq 0, x \in I = (0, 1))$. Moreover, the strain and the temperature are required to satisfy $\varepsilon > -1$ and $\theta > 0$, in the absence of the body force b and the external heat supply r, assuming that the material density ρ is equal to one, and taking in some considerations, the equations together with Cattaneo's law take the form

$$\begin{cases} u_{tt} - a u_{xx} + b\theta_x = \alpha_1 q q_x, \\ \theta_t + g q_x + d u_{tx} = \alpha_2 q q_t, \\ \tau q_t + q_x + k\theta_x = 0 \end{cases}$$

where

$$a = a(u_x, \theta, q), \ b = b(u_x, \theta, q), \ g = g(u_x, \theta, q), \ d = d(u_x, \theta, q),$$
$$\tau = \tau(u_x, \theta), \ \alpha_1 = \alpha_1(u_x, \theta), \ k = k(u_x, \theta), \ \alpha_2 = \alpha_2(u_x, \theta),$$

the authors established an exponential decay result for solutions with sufficiently small initial data and proved that the dissipation given by the heat conduction is strong enough to stabilize the system exponentially. This work has extended the result of Racke [235] to a more general situation.

We assume that S are C^3-functions satisfying

$$\frac{\partial S}{\partial u_x}(0, 0) = 1 > 0, \quad \frac{\partial S}{\partial \theta}(0, 0) \neq 0. \tag{2.1.6}$$

Concerning the kernel, we assume that $k_1(t) \in C^1(\mathbb{R}^+)$ and that $k_1(t)$ is a strongly positive definite kernel. Additionally, we assume that there exist positive constants $c_0 \leq c_1$, such that for all $t \geq 0$,

$$k_1(t) > 0, \quad k_1'(t) + c_0 k_1(t) \leq 0 \leq k_1'(t) + c_1 k_1(t). \tag{2.1.7}$$

To simplify notations, we shall introduce $-\frac{\partial S}{\partial \theta}(0, 0) = \alpha$ satisfying the product $\alpha\beta > 0$.

For the initial data, we assume that

$$\begin{cases} (u_0, u_1) \in \left(H^3(0, 1) \bigcap (H_0^1)\right) \times \left(H^2(0, 1) \bigcap H_0^1(0, 1)\right) \times H_0^1(0, 1), \\ (\theta_0, \theta_1) \in H^2(0, 1) \times H^1(0, 1), \\ (q_0, q_1) \in H^2(0, 1) \times H^1(0, 1), \end{cases} \tag{2.1.8}$$

with

$$\int_0^1 \theta_0(x)dx = 0, \quad u_2 = u_{tt}\mid_{t=0} = \frac{d}{dx}S(u_{0x}(x), \theta_0(x)). \tag{2.1.9}$$

We put $\|\cdot\| = \|\cdot\|_{L^2(0,1)}$, and use C (sometimes $C_1, C_2, \ldots$) to denote the generic positive constant independent of time $t > 0$.

Our main result of this chapter reads as follows.

Theorem 2.1.1 *Under assumptions (2.1.6)–(2.1.9), there exists a small constant $0 < \epsilon_0 < 1$ such that for any $\epsilon \in (0, \epsilon_0)$ and for any initial data $(u_0, u_1, \theta_0, q_0)$ satisfying*

$$\|u_0\|_{H^3}^2 + \| u_1 \|_{H^2}^2 + \| \theta_0 \|_{H^2}^2 + \| q_0 \|_{H^2}^2 < \epsilon, \tag{2.1.10}$$

problem (2.1.1)–(2.1.5) admits a unique global solution $(u(t), \theta(t), q(t))$ satisfying

$$u(t) \in \bigcap_{m=0}^{2} C^m([0, +\infty), H^{3-m}(0, 1) \cap H_0^1(0, 1)), \ (\partial_t^3 u)(t) \in C([0, +\infty), L^2(0, 1)), \tag{2.1.11}$$

$$(k_1 * \theta)(t), \theta(t) \in \bigcap_{m=0}^{1} C^m([0, +\infty), H^{2-m}(0, 1)), \tag{2.1.12}$$

$$(k_1 * \theta)(t), \theta(t) \in C^2([0, +\infty), L^2(0, 1)), \tag{2.1.13}$$

$$q(t) \in C^1([0, +\infty), L^2(0, 1)) \cap C([0, +\infty), H^1(0, 1)), \tag{2.1.14}$$

$$(k_1 * \partial_t^i \theta_x)(t), \ (k_1 * \partial_t^j \theta_{xx})(t) \in L^2([0, +\infty), L^2(0, 1)), \ i = 0, 1, 2; \ j = 0, 1, \tag{2.1.15}$$

$$\partial_t^i q(t), \ \partial_t^j q_x(t) \in L^2([0, +\infty), L^2(0, 1)), \ i = 0, 1, 2; \ j = 0, 1. \tag{2.1.16}$$

Moreover, the solution $(u(t), \theta(t), q(t))$ decays exponentially as $t \to +\infty$, i.e., there exists a large time $t_0 > 0$ such that as $t \geq t_0$,

$$\| u(t) \|_{H^3}^2 + \| u_t(t) \|_{H^2}^2 + \| \theta(t) \|_{H^2}^2 + \| q(t) \|_{H^2}^2 \leq Ce^{-C't}, \tag{2.1.17}$$

where C and C' are positive constants.

2.2 Global Existence and Exponential Stability

In this section, we shall prove Theorem 2.1.1. The main idea of the proof is that we first shall prove the local solutions, by a standard contraction mapping argument, we then can show that problem (2.1.1)–(2.1.5) admits a unique local solution $(u(t), \theta(t), q(t))$ such that

$$u(t) \in \bigcap_{m=0}^{2} C^m([0,T), H^{3-m}(0,1) \cap H_0^1(0,1)), \quad (\partial_t^3 u)(t) \in C([0,T), L^2(0,1)), \tag{2.2.1}$$

$$(k_1 * \theta)(t), \theta(t) \in \bigcap_{m=0}^{1} C^m([0,T), H^{2-m}(0,1)), \tag{2.2.2}$$

$$(k_1 * \theta)(t), \theta(t) \in C^2([0,T), L^2(0,1)), \tag{2.2.3}$$

$$(k_1 * \partial_t^i \theta_x)(t),\ (k_1 * \partial_t^j \theta_{xx})(t) \in L^2([0,T), L^2(0,1)),\ i = 0,1,2;\ j = 0,1, \tag{2.2.4}$$

$$\partial_t^i q(t),\ \partial_t^j q_x(t) \in L^2([0,T), L^2(0,1)),\ i = 0,1,2;\ j = 0,1, \tag{2.2.5}$$

$$q(t) \in C^1([0,T), L^2(0,1)) \cap C([0,T), H^1(0,1)), \tag{2.2.6}$$

where the constant $T > 0$ is the maximal existence interval of solutions, and second we shall use the contradiction argument to continue the local solutions in time. The next lemma concerns the property of a strongly positive definite kernel.

Lemma 2.2.1 *Assume that $\hat{k}(t) \in L^1(\mathbb{R}^+)$ is a strongly positive definite kernel satisfying $\hat{k}'(t) \in L^1(\mathbb{R}^+)$, then for any $y(t) \in L^1_{loc}(\mathbb{R}^+)$, it follows that*

$$\int_0^t \left|\hat{k} * y(\tau)\right|^2 d\tau \le \beta_0 k_2 \int_0^t y(\tau)\ \hat{k} * y(\tau) d\tau \tag{2.2.7}$$

where $k_2 = \left(\int_0^{+\infty} \left|\hat{k}(t)\right| dt\right)^2 + 4\left(\int_0^{+\infty} \left|\hat{k}'(t)\right| dt\right)^2$ and $\beta_0 > 0$ is a constant such that the function $\hat{k}(t) - \beta_0 e^{-t}$ is a positively definite kernel.

Proof Define

$$y_t(\tau) = \begin{cases} y(\tau), & 0 \le \tau \le t, \\ 0, & \text{otherwise.} \end{cases}$$

By the Plancherel identity and the fact that convolution is mapped into pointwise multiplication by the Fourier transform,

$$\begin{aligned} \int_0^t \left|\hat{k} * y(\tau)\right|^2 d\tau &\le \int_0^{+\infty} \left|\int_0^\tau \hat{k}(\tau - s) y_t(s) ds\right|^2 d\tau \\ &= \frac{1}{2\pi} \int_{-\infty}^{+\infty} \left|\tilde{\hat{k}}(w)\right|^2 |\tilde{y}_t(w)|^2 dw \end{aligned} \tag{2.2.8}$$

where

$$\left|\tilde{\hat{k}}(w)\right| = \left|\int_0^{+\infty} e^{-iwt}\hat{k}(t)dt\right| \le \int_0^{+\infty} |\hat{k}(t)|dt, \tag{2.2.9}$$

$$\left|w\tilde{\hat{k}}(w)\right| = \left|\int_0^{+\infty} (e^{-iwt} - 1)\hat{k}'(t)dt\right| \le 2\int_0^{+\infty} \left|\hat{k}'(t)\right| dt \tag{2.2.10}$$

and $\tilde{f}$ denotes the Fourier transform of f.
Square these two inequalities (2.2.9)–(2.2.10), and add (2.2.10) to (2.2.9) to get

$$\left|\tilde{\hat{k}}(w)\right|^2 \le \frac{k_2}{1+w^2}$$

which, combined with (2.2.8), yields

$$\begin{aligned}
&\int_0^t \left|\int_0^\tau \hat{k}(\tau - s)y(s)ds\right|^2 d\tau \\
&\le \frac{k_2}{2\pi}\int_{-\infty}^{+\infty} \frac{|\tilde{y}_t(w)|^2}{1+w^2}dw \\
&= \frac{k_2}{2}\int_{-\infty}^{+\infty} y_t(\tau)\int_{-\infty}^{+\infty} e^{-|\tau-s|}y_t(s)dsd\tau = k_2\int_0^t y(\tau)\int_0^\tau e^{-(\tau-s)}y(s)dsd\tau \\
&= k_2\int_0^t y(\tau)(e^{-\tau} * y)(\tau)d\tau \le k_2\beta_0\int_0^t y(\tau)(\hat{k} * y)(\tau)d\tau.
\end{aligned}$$

Thus (2.2.7) follows. □

From (2.1.3), we can deduce that

$$q = q_0 e^{-t} - \int_0^t e^{s-t}\theta_x ds. \tag{2.2.11}$$

By inserting (2.2.11) and (2.1.6) into (2.1.1)–(2.1.3), system (2.1.1)–(2.1.3) reduces to the system

$$\begin{cases}
u_{tt} - u_{xx} + \alpha\phi_x = f, & (2.2.12)\\
\theta_t - \tilde{k}_1 * \theta_{xx} + \beta u_{xt} = g, & (2.2.13)\\
q_t + q + k\theta_x = 0, & (2.2.14)
\end{cases}$$

where

$$\begin{cases}
f = \left[\dfrac{\partial S}{\partial u_x}(u_x,\theta) - 1\right]u_{xx} + \left[\dfrac{\partial S}{\partial \theta}(u_x,\theta) + \alpha\right]\theta_x, & (2.2.15)\\
g = -\gamma q_{0x}k_2 - k_1 * k_2 q_{0x}, \quad k_2 = e^{-t}, \quad \tilde{k}_1 = \gamma k k_2 + k k_1 * k_2. & (2.2.16)
\end{cases}$$

For simplicity, we put

$$\eta_1 = \frac{\partial S}{\partial u_x}(u_x, \theta) - 1, \quad \eta_2 = \frac{\partial S}{\partial \theta}(u_x, \theta) + \alpha. \tag{2.2.17}$$

From (2.1.7), we can derive that, if $\gamma > k_1(0)$, then there exists positive constants $c_2 \le c_3$, such that for all $t \ge 0$,

$$\tilde{k}_1(t) > 0, \quad \tilde{k}_1'(t) + c_2\tilde{k}_1(t) \le 0 \le \tilde{k}_1'(t) + c_3\tilde{k}_1(t). \tag{2.2.18}$$

If $\gamma \le k_1(0)$, from (2.1.7), there exists some time $t_0 > 0$, as $t \ge t_0$, we have $\gamma > k_1(t_0)$, so we can define $\hat{t} = t - t_0$, then we have for all $\hat{t} \ge 0$,

$$\tilde{k}_1(\hat{t}) > 0, \quad \tilde{k}_1'(\hat{t}) + c_2\tilde{k}_1(\hat{t}) \le 0 \le \tilde{k}_1'(\hat{t}) + c_3\tilde{k}_1(\hat{t}).$$

Thus without loss of generality, we may assume $\gamma > k_1(0)$.
It follows from (2.2.18) that the kernel $\tilde{k}_1(t)$ decays exponentially as time goes to infinity and satisfies

$$\tilde{k}_1(0)e^{-c_3 t} \le \tilde{k}_1(t) \le \tilde{k}_1(0)e^{-c_2 t}.$$

Thus, we can choose $\delta \in \delta_0 \equiv (0, \min(1, c_0/2, c_2/2))$ so small that for any $t \ge 0$,

$$\tilde{k}_1(0)e^{-(c_3/2)t} \le \hat{k}(t) := e^{\delta t}\tilde{k}_1(t) \le \tilde{k}_1(0)e^{-(c_2/2)t} \tag{2.2.19}$$

and for all $t \ge 0$,

$$\hat{k}(t) > 0, \quad \hat{k}'(t) + \frac{c_2}{2}\hat{k}(t) \le 0 \le \hat{k}'(t) + c_3\hat{k}(t). \tag{2.2.20}$$

Let us denote

$$v(x,t) = e^{\delta t}u(x,t), \ \phi(x,t) = e^{\delta t}\theta(x,t), \ p(x,t) = e^{\delta t}q(x,t). \tag{2.2.21}$$

Then the system (2.2.12)–(2.2.14) can be rewritten as

$$\begin{cases} v_{tt} - v_{xx} + \alpha\phi_x = F, & (2.2.22) \\ \phi_t - \hat{k} * \phi_{xx} + \beta v_{xt} = G, & (2.2.23) \\ p_t + (1-\delta)p + k\phi_x = 0, & (2.2.24) \\ v(0,t) = v(1,t) = p(0,t) = p(1,t) = 0, & (2.2.25) \\ v(x,0) = v_0, \ v_t(x,0) = v_1, \ \phi(x,0) = \phi_0, \ p(x,0) = p_0, & (2.2.26) \end{cases}$$

where

$$F(t) = fe^{\delta t} + 2\delta v_t - \delta^2 v, \;\; G(t) = ge^{\delta t} + \delta\phi + \delta\beta v_x, \tag{2.2.27}$$

$$\int_0^1 \phi_0(x)dx = \int_0^1 \theta_0(x)dx = 0. \tag{2.2.28}$$

We easily derive from (2.1.2), (2.1.10), and (2.2.28) that

$$\int_0^1 \phi(x,t)dx = \int_0^1 \theta(x,t)dx = 0. \tag{2.2.29}$$

To facilitate our analysis, let us introduce the linear problem

$$\begin{cases} V_{tt} - V_{xx} + \alpha\Phi_x = \mathcal{F}, & (2.2.30)\\ \Phi_t - \hat{k} * \Phi_{xx} + \beta V_{xt} = \mathcal{G}, & (2.2.31)\\ P_t + (1-\delta)P + k\Phi_x = 0, & (2.2.32)\\ V(0,t) = V(1,t) = P(0,t) = P(1,t) = 0, & (2.2.33)\\ V(x,0) = V_0, \;\; V_t(x,0) = V_1, \;\; \Phi(x,0) = \Phi_0, \;\; P(x,0) = P_0, & (2.2.34) \end{cases}$$

with

$$\int_0^1 \Phi_0(x)dx = 0. \tag{2.2.35}$$

It follows from (2.2.29) and (2.2.35) that when $(V, \Phi) = (v, \phi), (\mathcal{F}, \mathcal{G}) = (F, G)$ or $(V, \Phi) = (v_t, \phi_t), (\mathcal{F}, \mathcal{G}) = (F_t, G_t + \hat{k}(t)\phi_{0xx})$,

$$\int_0^1 \Phi(x,t)dx = 0, \quad for \;\; all\; t > 0. \tag{2.2.36}$$

In the sequel, we shall study the linearized system (2.2.30)–(2.2.32). To this end, we define the following energy functions

$$\begin{cases} E_1(t,V,\Phi) = \dfrac{1}{2}\displaystyle\int_0^1 (V_t^2 + V_x^2 + \alpha\beta^{-1}\Phi^2)dx, & (2.2.37)\\ E_2(t,V,\Phi) = \dfrac{1}{2}\displaystyle\int_0^1 (V_{tt}^2 + V_{tx}^2 + \alpha\beta^{-1}\Phi_t^2)dx, & (2.2.38)\\ E_3(t,V,\Phi) = \dfrac{1}{2}\displaystyle\int_0^1 (V_{tx}^2 + V_{xx}^2 + \alpha\beta^{-1}\Phi_x^2)dx. & (2.2.39) \end{cases}$$

Multiplying (2.2.25) and (2.2.26) by V_t and $\alpha\beta^{-1}\Phi$, respectively, and summing the results, we have

$$\frac{d}{dt}E_1(t, V, \Phi) = -\alpha\beta^{-1}\int_0^1 \Phi_x\hat{k} * \Phi_x dx + \int_0^1 (\mathcal{F}V_t + \alpha\beta^{-1}\mathcal{G}\Phi)dx. \quad (2.2.40)$$

Assuming regular initial data and noting that V_t and Φ_t satisfy the same boundary conditions, we get

$$\begin{aligned}\frac{d}{dt}E_2(t, V, \Phi) &= -\alpha\beta^{-1}\int_0^1 \Phi_{tx}\hat{k} * \Phi_{tx}dx - \alpha\beta^{-1}\hat{k}(t)\int_0^1 \Phi_{0x}\Phi_{tx}dx \\ &\quad + \int_0^1 (\mathcal{F}_t V_{tt} + \alpha\beta^{-1}\mathcal{G}_t\Phi_t)dx \\ &= -\alpha\beta^{-1}\int_0^1 \Phi_{tx}\hat{k} * \Phi_{tx}dx - \alpha\beta^{-1}\frac{d}{dt}\Big(\hat{k}(t)\int_0^1 \Phi_{0x}\Phi_{tx}dx\Big) \\ &\quad + \alpha\beta^{-1}\frac{d}{dt}\hat{k}'(t)\int_0^1 \Phi_{0x}\Phi_{tx}dx + \int_0^1 (\mathcal{F}_t V_{tt} + \alpha\beta^{-1}\mathcal{G}_t\Phi_t)dx.\end{aligned} \quad (2.2.41)$$

Similarly, multiplying (2.2.30) and (2.2.31) by V_{xxt} and $\alpha\beta^{-1}\Phi_{xx}$, respectively, and summing the results, we have

$$\begin{aligned}\frac{d}{dt}E_3(t, V, \Phi) &= -\alpha\beta^{-1}\int_0^1 \Phi_{xx}\hat{k} * \Phi_{xx}dx + \int_0^1 (\mathcal{F}V_{xxt} + \alpha\beta^{-1}\mathcal{G}\Phi_{xx})dx \\ &= -\alpha\beta^{-1}\int_0^1 \Phi_{xx}\hat{k} * \Phi_{xx}dx - \frac{d}{dt}\int_0^1 \mathcal{F}V_{xx}dx \\ &\quad + \int_0^1 (\mathcal{F}_t V_{xt} + \alpha\beta^{-1}\mathcal{G}\Phi_{xx})dx.\end{aligned} \quad (2.2.42)$$

Now we introduce the following functions:

$$\begin{cases} E_4(t, V, \Phi) = -\int_0^1\int_0^x \Phi_t dy V_{tt}dx, \\ E_5(t, V, \Phi) = \int_0^1 V_{tx}\Phi dx, \quad E_6(t, V, \Phi) = \int_0^1 V_{tx}V_x dx, \\ E_7(t, V, \Phi) = -\int_0^1 \Phi_t\hat{k}\Phi_t dx, \quad E_8(t, V, \Phi) = -\int_0^1 \Phi_x\hat{k}\Phi_x dx. \end{cases}$$

Thus, integrating (2.2.31) over $(0, x)$, using the boundary conditions and (2.2.32), we derive

$$\int_0^x \Phi_t dy - \hat{k} * \Phi_x + \beta V_t = \int_0^x \mathcal{G}dy. \quad (2.2.43)$$

By (2.2.30) and (2.2.43), we easily get

$$E_4(t, V, \Phi) \le -\frac{\beta}{2} \parallel V_{tt} \parallel^2 + \frac{\beta}{8} \parallel V_{tx} \parallel^2 + \frac{1}{\beta}\Big(k^2(0) \parallel \Phi_x \parallel^2 + \parallel \hat{k}' * \Phi_x \parallel^2\Big)$$
$$+ (\alpha + \frac{2}{\beta}) \parallel \Phi_t \parallel^2 + \int_0^1 \Big(\int_0^x \mathcal{G}_t dy V_{tt} + \int_0^x \Phi_t dy \mathcal{F}_t\Big) dx. \tag{2.2.44}$$

Now we define

$$n(t, V, \Phi) = \int_0^1 \Big(V_{tt}^2 + V_{tx}^2 + V_{xx}^2 + \Phi_t^2 + \Phi_x^2\Big)(t) dx$$

and

$$L(t, V, \Phi) = N\Big(E_1(t, V, \Phi) + E_2(t, V, \Phi) + E_3(t, V, \Phi) + \alpha\beta^{-1}\hat{k}\int_0^1 \Phi_{0x}\Phi_x dx\Big)$$
$$+ E_4(t, V, \Phi) + E_5(t, V, \Phi) + \frac{\beta}{4}E_6(t, V, \Phi) + a_1 E_7(t, V, \Phi) + a_2 E_8(t, V, \Phi),$$

where $N > 0$ is a parameter sufficiently large and

$$a_1 = \frac{4}{\tilde{k}_1(0)}\Big(\alpha + \frac{2}{\beta}\Big), \quad a_2 = \frac{4}{\tilde{k}_1(0)}\Big(\alpha + \frac{4 + \tilde{k}_1^2(0)}{\beta} + \frac{\alpha^2\beta}{8} + a_1\Big).$$

Under the above notations, we can derive the following lemma.

Lemma 2.2.2 *There exist positive constants $\beta_1, \beta_2, \beta_3, C_3, C_4$ and sufficiently large constant N such that $L(t, V, \Phi)$ satisfies the following inequalities:*

$$\frac{d}{dt}L(t, V, \Phi) \le -C_3 n(t, V, \Phi) + C_4\Big(\parallel \hat{k} * \Phi_x \parallel^2 + \parallel \hat{k} * \Phi_{xt} \parallel^2 + \parallel \hat{k} * \Phi_{xx} \parallel^2 \Big)$$
$$- \alpha N\beta^{-1}\int_0^1 (\Phi_x \hat{k} * \Phi_x + \Phi_{tx}\hat{k} * \Phi_{tx} + \Phi_{xx}\hat{k} * \Phi_{xx}) dx$$
$$+ R(t, V, \Phi), \tag{2.2.45}$$

$$L(t, V, \Phi) \le \beta_2\Big(n(t, V, \Phi) + \parallel \hat{k} * \Phi_x \parallel^2 + \parallel \hat{k} * \Phi_t \parallel^2 + \hat{k}^2(t) \parallel \Phi_{0x} \parallel^2\Big), \tag{2.2.46}$$

$$L(t, V, \Phi) \ge \beta_1 n(t, V, \Phi) - \beta_3\Big(\parallel \hat{k} * \Phi_x \parallel^2 + \parallel \hat{k} * \Phi_t \parallel^2 + \hat{k}^2(t) \parallel \Phi_{0x} \parallel^2 \Big), \tag{2.2.47}$$

where

$$\begin{aligned}R(t, V, \Phi) &= N\int_0^1 \Big(\mathcal{F}V_t + \alpha\beta^{-1}\mathcal{G}\Phi + \mathcal{F}_t V_{tt} + \alpha\beta^{-1}\mathcal{G}_t\Phi_t + \mathcal{F}_t V_{xx} - \alpha\beta^{-1}\mathcal{G}\Phi_{xx}\Big)\,dx\\ &\quad -N\frac{d}{dt}\int_0^1 \mathcal{F}V_{xx}dx + \alpha N\beta^{-1}\hat{k}'(t)\int_0^1 \Phi_{0x}\Phi_x dx - a_1\int_0^1 \mathcal{G}_t\hat{k}*\Phi_t dx\\ &\quad +\int_0^1 \Big(\int_0^x \mathcal{G}_t dy V_{tt} + \int_0^x \Phi_t dy \mathcal{F}_t\Big)dx + a_2\int_0^1 \mathcal{G}\hat{k}*\Phi_{xx}dx\\ &\quad +\int_0^1 \Big(V_{tx}\mathcal{G} - \mathcal{F}\Phi_x - \frac{\beta}{4}\mathcal{F}V_{xx}\Big)dx. \end{aligned} \tag{2.2.48}$$

Proof By (2.2.30)–(2.2.31) and integration by parts, we get

$$\begin{aligned}\frac{d}{dt}E_5(t, V, \Phi) &= -\beta \parallel V_{tx} \parallel^2 + \alpha \parallel \Phi_x \parallel^2 - \int_0^1 (V_{xx} + \mathcal{F})\Phi_x dx\\ &\quad + \int_0^1 V_{tx}(\hat{k} * \Phi_{xx} + \mathcal{G})dx\\ &\le -\frac{\beta}{2} \parallel V_{tx} \parallel^2 + \frac{\beta}{16} \parallel V_{xx} \parallel^2 + \Big(\alpha + \frac{4}{\beta}\Big) \parallel \Phi_x \parallel^2\\ &\quad + \frac{1}{2\beta} \parallel \hat{k} * \Phi_{xx} \parallel^2 + \int_0^1 (V_{tx}\mathcal{G} - \mathcal{F}\Phi_x)dx \end{aligned} \tag{2.2.49}$$

and

$$\begin{aligned}\frac{d}{dt}E_6(t, V, \Phi) &= \parallel V_{tx} \parallel^2 - \parallel V_{xx} \parallel^2 + \alpha\int_0^1 \Phi_x V_{xx}dx - \int_0^1 \mathcal{F}V_{xx}dx\\ &\le -\frac{1}{2} \parallel V_{xx} \parallel^2 + \frac{\alpha^2}{2} \parallel \Phi_x \parallel^2 + \parallel V_{tx} \parallel^2 - \int_0^1 \mathcal{F}V_{xx}dx. \end{aligned} \tag{2.2.50}$$

Thus it follows from (2.2.44) and (2.2.49)–(2.2.50) that

$$\begin{aligned}&\frac{d}{dt}\Big(E_4(t, V, \Phi) + E_5(t, V, \Phi) + \frac{\beta}{4}E_6(t, V, \Phi)\Big)\\ &\le -\frac{\beta}{2} \parallel V_{tt} \parallel^2 - \frac{\beta}{16} \parallel V_{xx} \parallel^2 - \frac{\beta}{8} \parallel V_{tx} \parallel^2 + \Big(\alpha + \frac{2}{\beta}\Big) \parallel \Phi_t \parallel^2\\ &+ \Big(\alpha + \frac{4}{\beta} + \frac{\alpha^2\beta}{8} + \frac{\tilde{k}_1^2(0)}{\beta}\Big) \parallel \Phi_x \parallel^2 + \frac{1}{\beta} \parallel \hat{k}' * \Phi_x \parallel^2 + \frac{1}{2\beta} \parallel \hat{k} * \Phi_{xx} \parallel^2\\ &+ \int_0^1 \Big(\int_0^x \mathcal{G}_t dy V_{tt} + \int_0^x \Phi_t dy \mathcal{F}_t\Big)dx\\ &+ \int_0^1 \Big(V_{tx}\mathcal{G} - \mathcal{F}\Phi_x - \frac{\beta}{4}\mathcal{F}V_{xx}\Big)dx. \end{aligned} \tag{2.2.51}$$

On the other hand, differentiating (2.2.31) with respect to t, multiplying the resulting equation by $\hat{k} * \Phi_t$ and integrating it by parts, we deduce

$$\begin{aligned}\frac{d}{dt}E_7(t, V, \Phi) &= -\tilde{k}_1(0) \parallel \Phi_t \parallel^2 + \int_0^1 \left(\Phi_t \hat{k}' * \Phi_t + \tilde{k}_1(0)\Phi_x \hat{k} * \Phi_{tx} + \hat{k}' * \Phi_x \hat{k} * \Phi_{tx}\right) dx \\ &\quad - \int_0^1 (\beta V_{tt} \hat{k} * \Phi_{tx} - \mathcal{G}_t \hat{k} * \Phi_t) dx \\ &\le -\frac{\tilde{k}_1(0)}{2} \parallel \Phi_t \parallel^2 + \frac{\beta}{4a_1} \parallel V_{tt} \parallel^2 + \parallel \Phi_x \parallel^2 + \frac{1}{2\tilde{k}_1(0)} \parallel \hat{k}' * \Phi_t \parallel^2 \\ &\quad + \frac{\tilde{k}_1^2(0)}{4} \parallel \hat{k} * \Phi_{tx} \parallel^2 + \frac{1}{2}\left(\parallel \hat{k}' * \Phi_x \parallel^2 + \parallel \hat{k} * \Phi_{tx} \parallel^2\right) \\ &\quad + \beta a_1 \parallel \hat{k} * \Phi_{tx} \parallel^2 - \int_0^1 \mathcal{G}_t \hat{k} * \Phi_t dx. \end{aligned} \tag{2.2.52}$$

Similarly, differentiating (2.2.31) with respect to x, multiplying the resulting equation by $\hat{k} * \Phi_x$ and integrating it by parts, we infer

$$\begin{aligned}\frac{d}{dt}E_8(t, V, \Phi) &= -\tilde{k}_1(0) \parallel \Phi_x \parallel^2 + \parallel \hat{k} * \Phi_{xx} \parallel^2 + \int_0^1 \mathcal{G} \hat{k} * \Phi_{xx} dx \\ &\quad - \int_0^1 \left(\beta V_{tx} \hat{k} * \Phi_{xx} + \Phi_x \hat{k}' * \Phi_x\right) dx \\ &\le -\frac{\tilde{k}_1(0)}{2} \parallel \Phi_x \parallel^2 + \frac{\beta}{4a_2} \parallel V_{tx} \parallel^2 + \frac{1}{2\tilde{k}_1(0)} \parallel \hat{k}' * \Phi_x \parallel^2 \\ &\quad + (1 + \beta a_2) \parallel \hat{k} * \Phi_{xx} \parallel^2 + \int_0^1 \mathcal{G} \hat{k} * \Phi_{xx} dx. \end{aligned} \tag{2.2.53}$$

Combining (2.2.52) and (2.2.53) with (2.2.51) gives

$$\begin{aligned}&\frac{d}{dt}\left(E_4(t, V, \Phi) + E_5(t, V, \Phi) + \frac{\beta}{4}E_6(t, V, \Phi) + a_1 E_7(t, V, \Phi) + a_2 E_8(t, V, \Phi)\right) \\ &\le -C_3 n(t, V, \Phi) + \left(\frac{1}{\beta} + \frac{a_1}{2} + \frac{a_2}{2\tilde{k}_1(0)}\right) \parallel \hat{k}' * \Phi_x \parallel^2 \\ &\quad + \left[\frac{1}{\beta} + (1 + \beta a_2) a_2\right] \parallel \hat{k} * \Phi_{xx} \parallel^2 + \frac{a_1}{2\tilde{k}_1(0)} \parallel \hat{k}' * \Phi_t \parallel^2 \\ &\quad + \left(\tilde{k}_1^2(0)\frac{a_1}{4} + \frac{a_1}{2} + \beta a_1^2\right) \parallel \hat{k} * \Phi_{tx} \parallel^2 + R_1(t, V, \Phi), \end{aligned} \tag{2.2.54}$$

where $C_3 = \min\left(\beta/16, \tilde{k}_1(0)a_1/4, \tilde{k}_1(0)a_2/4\right)$. In view of (2.2.20), (2.2.36) and Poincare's inequality, we have

$$\parallel \hat{k}' * \Phi_t \parallel \le \parallel \hat{k}' * \Phi_{tx} \parallel \le C \parallel \hat{k} * \Phi_{tx} \parallel . \tag{2.2.55}$$

Thus it follows from (2.2.40)–(2.2.42) and (2.2.54)–(2.2.55) that (2.2.45) holds. From the definition of $L(t, V, \Phi)$, we easily know that there exist constants $\beta_1, \beta_2, \beta_3 > 0$ and a sufficiently large constant N such that (2.2.46) and (2.2.47) hold. The proof is complete. □

Now we define

$$\mathcal{M}(t, V, \Phi) = n(t, v, \phi) + n(t, v_t, \phi_t) + \| \phi_{xx}(t) \|^2 .$$

Differentiating (2.2.23) with respect to t, we arrive at

$$\phi_{tt} - \tilde{k}_1(0)\phi_{xx} - \hat{k} * \phi_{xx} + \beta v_{ttx} = G_t,$$

which, combined with (2.2.20), (2.2.23) and (2.2.27), yields

$$\begin{aligned} \| \phi_{xx} \|^2 &\le C\Big(\| \phi_{tt} \|^2 + \| v_{ttx} \|^2 + \| \phi_t \|^2 + \| v_{tx} \|^2 + \| G \|^2 + \| G_t \|^2 \Big) \\ &\le C_4\Big(n(t, v, \phi) + n(t, v_t, \phi_t)\Big) + C_5\Big((k_2(t))^2 + ((k_1 * k_2)(t))^2\Big) \| p_{0x} \|^2 . \end{aligned} \tag{2.2.56}$$

Thus,

$$\begin{aligned} n(t, v, \phi) + n(t, v_t, \phi_t) \le \mathcal{M}(t, v, \phi) &\le C_6\Big(n(t, v, \phi) + n(t, v_t, \phi_t)\Big) \\ &+ C_5\Big((k_2(t))^2 + ((k_1 * k_2)(t))^2\Big) \| p_{0x} \|^2 . \end{aligned} \tag{2.2.57}$$

By the smallness condition (2.1.16) of initial data, there is a constant $\alpha_1 > 1$, independent of δ, such that

$$\mathcal{M}(0, u, \theta, q) < \alpha_1 \epsilon^2. \tag{2.2.58}$$

Using equations (2.2.18)–(2.2.21), there exists a constant $\alpha_2 > 1$, independent of δ, such that

$$n(0, v, \phi) + n(0, v_t, \phi_t) \le \mathcal{M}(0, v, \phi) \le \alpha_2 \mathcal{M}(0, u, \theta) < \alpha_1\alpha_2\epsilon^2. \tag{2.2.59}$$

We derive from (2.1.7), (2.2.16), (2.2.19)–(2.2.20) and (2.2.59) that there exists a constant $\eta_0 > 0$, independent of δ, such that

$$\begin{aligned} \int_0^{+\infty} &\Big\{\lambda_3(\hat{k}'(t))^2\Big(\| \phi_{0x} \|^2 + \| \phi_{1x} \|^2 \Big) + \Big[\lambda_1(\hat{k}'(t))^2 + \lambda_2(\hat{k}(t))^2\Big] \| \phi_{0xx} \|^2 \\ &+ \lambda_4\Big((k_2(t))^2 + ((k_1 * k_2)(t))^2 + ((k_1 * k_2)'(t))^2\Big) \| p_{0x} \|^2 \Big\}dt < \eta_0\epsilon^2, \end{aligned} \tag{2.2.60}$$

where

$$\lambda_1 = 1 + \frac{2C_6}{C_3} + \frac{4N^2\alpha^2 C_6}{C_3\beta^2}, \quad \lambda_2 = \frac{2C_6}{C_3} + \frac{2N^2\alpha^2 C_6}{C_3\beta^2},$$
$$\lambda_3 = \frac{2N^2\alpha^2 C_6}{C_3\beta^2}, \quad \lambda_4 = \lambda_2 + NC_5. \tag{2.2.61}$$

Using the continuity of the solutions, it follows that there exist constants $\alpha_0 > 0$ and $t_0 \in [0, T)$ such that

$$\mathcal{M}(t, v, \phi) \le \alpha_0 \epsilon^2, \quad for \ \ all\ t \in [0, t_0). \tag{2.2.62}$$

Now we define

$$t_1 = \sup\left\{\tau_1 > 0;\ \mathcal{M}(t, v, \phi) \le \alpha_0\epsilon^2 \ \text{ in } \ [0, \tau_1)\right\}. \tag{2.2.63}$$

By Sobolev's embedding theorem and (2.2.63), we obtain that for any $(x, t) \in [0, 1]\times [0, t_1)$,

$$| v_x(x,t) | + | \phi(x,t) | + | \phi_x(x,t) | + | \phi_t(x,t) | \le C_7\epsilon, \tag{2.2.64}$$

which implies that for any $(x, t) \in [0, 1] \times [0, t_1)$,

$$| u_x(x,t) | + | \theta(x,t) | + | \theta_x(x,t) | + | \theta_t(x,t) | \le C_8\epsilon e^{-\delta t}. \tag{2.2.65}$$

Thus, if ϵ is small enough, we have that for any $(x, t) \in [0, 1] \times [0, t_1)$,

$$| u_x(x,t) | < \rho_0. \tag{2.2.66}$$

Define

$$\nu = \sup_{|x|+|y|\le\rho_0} \{|\partial^\rho \eta_i|;\ i = 1, 2;\ 0 \le |\rho| \le 3\},$$

where ∂^ρ denotes the partial derivatives of order $|\rho|$. Recalling the definitions of $\eta_i (i = 1, 2)$ and using the above inequalities, we deduce

$$| \eta_i | \le C_9\epsilon, \quad i = 1, 2, \tag{2.2.67}$$

with $C_9 = C_9(\nu) > 0$ being a constant. By (2.2.63)–(2.2.67), we easily derive that for any $(x, t) \in [0, 1] \times [0, t_1)$,

$$| v_t(x,t) | + | v_{tx}(x,t) | + | v_{tt}(x,t) | \le C_{10}\epsilon,$$

which, together with (2.2.19), (2.2.21), implies that for any $(x,t)\in[0,1]\times[0,t_1)$,

$$| u_t(x,t) | + | u_{tx}(x,t) | + | u_{tt}(x,t) | \le C_{11}e^{-\delta t}\epsilon. \tag{2.2.68}$$

By (2.2.22), (2.2.64), and (2.2.67)–(2.2.68), we get

$$| v_{xx}(x,t) | \le C\epsilon + C\epsilon \, | v_{xx}(x,t) |,$$

which gives

$$| v_{xx}(x,t) | \le C_{12}\epsilon, \quad | u_{xx}(x,t) | \le C_{12}\epsilon e^{-\delta t}, \ \textit{for all } (x,t)\in[0,1]\times[0,t_1). \tag{2.2.69}$$

Similarly, differentiating (2.2.22) with respect to x, we conclude

$$\| v_{xxx}(t) \|^2 \le C\Big[\mathcal{M}(t,v,\phi) + \| F_x \|^2\Big] \le C\mathcal{M}(t,v,\phi) + C\epsilon^2 \| v_{xxx}(t) \|^2,$$

which gives that for any $t\in[0,t_1)$,

$$\| v_{xxx}(t) \|^2 \le C\mathcal{M}(t,v,\phi) \le C_{13}\epsilon^2, \quad \| u_{xxx}(t) \|^2 \le C\mathcal{M}(t,v,\phi) \le C_{13}\epsilon^2 e^{-\delta t} \tag{2.2.70}$$

provided that ϵ is small enough.

Lemma 2.2.3 *Under the same assumptions as in Theorem 2.1.1, the following inequalities hold for any $t\in[0,t_1)$:*

$$\int_0^1 \mathcal{F}_t V_{tt} dx \le C(\delta+\epsilon)\mathcal{M}(t,v,\phi) - \frac{1}{2}\frac{d}{dt}\int_0^1 \eta_1 V_{tx}^2 dx, \tag{2.2.71}$$

$$\int_0^1 \mathcal{F}_t V_{xx} dx \le C(\delta+\epsilon)\mathcal{M}(t,v,\phi) + \frac{1}{2}\frac{d}{dt}\int_0^1 \eta_1 V_{xx}^2 dx, \tag{2.2.72}$$

$$\int_0^1 \mathcal{F} V_t dx \le C(\delta+\epsilon)\mathcal{M}(t,v,\phi), \tag{2.2.73}$$

$$\int_0^1 \mathcal{F} V_{xx} dx \le C(\delta+\epsilon)\mathcal{M}(t,v,\phi), \tag{2.2.74}$$

$$\int_0^1 \mathcal{F}^2 dx \le C(\delta+\epsilon)\mathcal{M}(t,v,\phi). \tag{2.2.75}$$

Proof We only consider the case of $(V,\Phi)=(v_t,\phi_t)$ and $\mathcal{F}=F_t$ to prove (2.2.71). The case of $(V,\Phi)=(v,\phi)$ and $\mathcal{F}=F$ is simple. By (2.2.15) and noting that

$$\begin{cases} F_{tt} = f_{tt}e^{\delta t} + 2\delta f_t e^{\delta t} + \delta^2 f e^{\delta t} + 2\delta v_{ttt} - \delta^2 v_t, \\ f_{tt} = \eta_{1tt}u_{xx} + 2\eta_{1t}u_{xxt} + \eta_1 u_{xxtt} + \eta_{2tt}\theta_x + 2\eta_{2t}\theta_{xt} + \eta_2\theta_{xtt}, \\ \eta_{1tt} = (u_{xt},\theta_t)\mathcal{H}_{\eta_1}(u_{xt},\theta_t)^\tau + \nabla\eta_1\cdot(u_{xtt},\theta_{tt}), \ \eta_{1t} = \nabla\eta_1\cdot(u_{xt},\theta_t), \end{cases}$$

we have

$$e^{\delta t} \parallel f_t \parallel \le C(\epsilon + \delta)\Big(\parallel v_{xx} \parallel + \parallel v_{txx} \parallel + \parallel \phi_x \parallel + \parallel \phi_{tx} \parallel \Big).$$

Here we only estimate the typical term in $\int_0^1 f_{tt} v_{ttt} e^{\delta t} dx$, that is, $\int_0^1 \eta_1 u_{xxtt} v_{ttt} e^{\delta t} dx$. Using (2.2.65) and (2.2.67)–(2.2.70), the other terms in $\int_0^1 f_{tt} v_{ttt} e^{\delta t} dx$ can be controlled by $C(\epsilon + \delta)\mathcal{M}(t, v, \phi)$ in the same way. Noting that

$$u_{xxtt} e^{\delta t} = v_{xxtt} - 2\delta v_{xxt} + \delta^2 v_{xx},$$
$$\parallel v_{ttt}(t) \parallel^2 \le C \left(\parallel v_{txx}(t) \parallel^2 + \parallel v_{xx}(t) \parallel^2 + \parallel \phi_x(t) \parallel^2 + \parallel \phi_{tx}(t) \parallel^2\right)$$

and using the integration by parts, we arrive at

$$\int_0^1 \eta_1 u_{xxtt} v_{ttt} e^{\delta t} dx \le C(\epsilon + \delta)\mathcal{M}(t, v, \phi) - \frac{1}{2}\frac{d}{dt}\int_0^1 \eta_1 v_{ttx}^2 dx.$$

Thus estimate (2.2.71) is valid. Similarly, we can prove estimates (2.2.72)–(2.2.75). The proof is complete. □

By Lemma 2.2.3, we can obtain the next two lemmas.

Lemma 2.2.4 *Under the same assumptions as in Theorem 2.1.1, the following inequalities hold for any* $t \in [0, t_1)$:

$$\begin{aligned} R(t, v, \phi) \le\ & C(\epsilon + \delta)\mathcal{M}(t, v, \phi) + C\delta(\parallel \hat{k} * \phi_{tx} \parallel^2 + \parallel \hat{k} * \phi_{xx} \parallel^2) \\ & + \frac{C_3}{8C_6} \parallel \phi_x \parallel^2 + \frac{2N^2\alpha^2 C_6}{C_3\beta^2} (\hat{k}'(t))^2 \parallel \phi_{0x} \parallel^2 \\ & + \frac{2N^2\alpha^2 C_6}{C_3\beta^2} \left((k_2(t))^2 + ((k_1 * k_2)(t))^2\right) \parallel p_{0x} \parallel^2 - N\frac{d}{dt}\int_0^1 F v_{xx} dx, \end{aligned}$$

$$N\int_0^1 F v_{xx} dx \le C(\epsilon + \delta)\mathcal{M}(t, v, \phi).$$

Lemma 2.2.5 *Under the same assumptions as in Theorem 2.1.1, the following inequalities hold for any* $t \in [0, t_1)$:

$$\begin{aligned} R(t, v_t, \phi_t)\ \le\ & C(\epsilon + \delta)\mathcal{M}(t, v, \phi) + (C\delta + \frac{a_1}{4}) \parallel \hat{k} * \phi_{tx} \parallel^2 + C\delta \parallel \hat{k} * \phi_{txx} \parallel^2 \\ & + \frac{C_3}{8C_6}\Big(\parallel v_{ttt} \parallel^2 + \parallel v_{ttx} \parallel^2 + \parallel \phi_t \parallel^2 + \parallel \phi_{tt} \parallel^2 + \parallel \phi_{tx} \parallel^2 + \parallel \phi_{xx} \parallel^2\Big) \\ & + \Big(\lambda_1(\hat{k}'(t))^2 + \lambda_2(\hat{k}(t))^2\Big) \parallel \phi_{0xx} \parallel^2 + \lambda_3(\hat{k}'(t))^2 \parallel \phi_{1x} \parallel^2 \\ & + \lambda_4\Big((k_2(t))^2 + ((k_1 * k_2)(t))^2 + ((k_1 * k_2)'(t))^2\Big) \parallel p_{0x} \parallel^2 \\ & + \frac{d}{dt}\int_0^1 \left[\frac{N}{2}(\eta_1 v_{txx}^2 - v_{ttx}^2) - N\alpha\beta^{-1}\hat{k}(t)\phi_{0xx}\phi_{xx} - NF_t v_{txx}\right] dx, \quad (2.2.76) \end{aligned}$$

$$N\int_0^1 F_t dx \le C(\epsilon+\delta)\mathcal{M}(t,v,\phi) - N\alpha\beta^{-1}\hat{k}(t)\int_0^1 \phi_{0xx}\phi_{xx}dx$$
$$\le \frac{\beta_1}{2}\Big(n(t,v,\phi)+n(t,v_t,\phi_t)\Big) + \frac{N^2\alpha^2 C_4}{2\beta_1\beta_2}\hat{k}^2(t)\parallel \phi_{0xx}\parallel^2. \quad (2.2.77)$$

Let us introduce the following function

$$\Upsilon_1(t) = L(t,v,\phi) + L(t,v_t,\phi_t) + N\int_0^1 (Fv_{xx}+F_t v_{txx})dx$$
$$+\frac{N}{2}\int_0^1 \eta_1(v_{ttx}^2 - v_{txx}^2)dx + N\alpha\beta^{-1}\hat{k}(t)\int_0^1 \phi_{0xx}\phi_{xx}dx.$$

Then it follows from (2.2.57), (2.2.65)–(2.2.70) and Lemmas 2.2.2–2.2.5. that if $\epsilon+\delta$ is small enough,

$$\Upsilon_1(t) \le \Big(\beta_2+\frac{\beta_1}{2}\Big)\Big(n(t,v,\phi)+n(t,v_t,\phi_t)\Big) + C(\epsilon+\delta)\mathcal{M}(t,v,\phi)$$
$$+\frac{N^2\alpha^2C_4}{2\beta_1\beta_2}\hat{k}^2(t)\parallel\phi_{0xx}\parallel^2 + \beta_2\Big(\parallel\hat{k}*\phi_t\parallel^2+\parallel\hat{k}*\phi_{tt}\parallel^2+\parallel\hat{k}*\phi_x\parallel^2$$
$$+\parallel\hat{k}*\phi_{tx}\parallel^2+\hat{k}^2(t)\parallel\phi_{0x}\parallel^2+\hat{k}^2(t)\parallel\phi_{1x}\parallel^2\Big)$$
$$\le \frac{(\beta_2+\beta_1)}{2}[n(t,v,\phi)+n(t,v_t,\phi_t)] \quad (2.2.78)$$
$$+C_5C(\epsilon+\delta)\Big((k_2(t))^2+((k_1*k_2)(t))^2\Big)\parallel p_{0x}\parallel^2$$
$$+\Big(\beta_2\parallel\phi_{0x}\parallel^2+\beta_2\parallel\phi_{1x}\parallel^2+\frac{N^2\alpha^2C_4}{2\beta_1\beta_2}\parallel\phi_{0xx}\parallel^2\Big)\hat{k}^2(t)$$
$$+\beta_2(\parallel\hat{k}*\phi_t\parallel^2+\parallel\hat{k}*\phi_{tt}\parallel^2+\parallel\hat{k}*\phi_x\parallel^2+\parallel\hat{k}*\phi_{tx}\parallel^2) \quad (2.2.79)$$

and

$$\Upsilon_1(t) \ge \frac{\beta_1}{2}[n(t,v,\phi)+n(t,v_t,\phi_t)] - C(\epsilon+\delta)\mathcal{M}(t,v,\phi)$$
$$-\frac{N^2\alpha^2C_4}{2\beta_1\beta_2}\hat{k}^2(t)\parallel\phi_{0xx}\parallel^2 - \beta_3\Big(\parallel\hat{k}*\phi_t\parallel^2+\parallel\hat{k}*\phi_{tt}\parallel^2$$
$$+\parallel\hat{k}*\phi_x\parallel^2+\parallel\hat{k}*\phi_{tx}\parallel^2+\hat{k}^2(t)\parallel\phi_{0x}\parallel^2+\hat{k}^2(t)\parallel\phi_{1x}\parallel^2\Big)$$
$$\ge \frac{\beta_1}{4}[n(t,v,\phi)+n(t,v_t,\phi_t)]$$
$$-C_5C(\epsilon+\delta)\Big((k_2(t))^2+((k_1*k_2)(t))^2\Big)\parallel p_{0x}\parallel^2$$
$$-\Big(\beta_3\parallel\phi_{0x}\parallel^2+\beta_2\parallel\phi_{1x}\parallel^2+\frac{N^2\alpha^2C_4}{2\beta_1\beta_2}\parallel\phi_{0xx}\parallel^2\Big)\hat{k}^2(t)$$
$$-\beta_3(\parallel\hat{k}*\phi_t\parallel^2+\parallel\hat{k}*\phi_{tt}\parallel^2+\parallel\hat{k}*\phi_x\parallel^2+\parallel\hat{k}*\phi_{tx}\parallel^2). \quad (2.2.80)$$

Define

$$\begin{aligned}\Upsilon(t) = \Upsilon_1(t) + C_5C(\epsilon+\delta)\Big((k_2(t))^2 + ((k_1*k_2)(t))^2\Big) \parallel p_{0x} \parallel^2 \\ + \Big(\beta_3 \parallel \phi_{0x} \parallel^2 + \beta_2 \parallel \phi_{1x} \parallel^2 + \frac{N^2\alpha^2 C_4}{2\beta_1\beta_2} \parallel \phi_{0xx} \parallel^2\Big)\hat{k}^2(t) \\ + \beta_3\Big(\parallel \hat{k}*\phi_t \parallel^2 + \parallel \hat{k}*\phi_{tt} \parallel^2 + \parallel \hat{k}*\phi_x \parallel^2 + \parallel \hat{k}*\phi_{tx} \parallel^2 \Big). \quad (2.2.81)\end{aligned}$$

Then it follows from (2.2.80), (2.2.20) and (2.2.57) that if $\epsilon+\delta$ is small enough, then we have

$$\Upsilon(t) \geq \frac{\beta_1(n(t,v,\phi)+n(t,v_t,\phi_t))}{4} \geq \frac{\beta_1}{4C_6}\mathcal{M}(t,v,\phi), \quad (2.2.82)$$

$$\begin{aligned}\frac{d}{dt}\Upsilon(t) &\leq \frac{d}{dt}\Upsilon_1(t) + \frac{C_3}{8}(n(t,v,\phi)+n(t,v_t,\phi_t)) \\ &\quad + C_{14}\beta_3\Big(\parallel \hat{k}*\phi_{tx} \parallel^2 + \parallel \hat{k}*\phi_{ttx} \parallel^2 + \parallel \hat{k}*\phi_x \parallel^2 \Big). \quad (2.2.83)\end{aligned}$$

Proof of Theorem 2.1.1 We shall assume that the initial data belong to $H^4(0,1)$. Our result will follow the standard density argument. By virtue of Lemmas 2.2.4 and 2.2.5, we easily obtain

$$\begin{aligned}\frac{d}{dt}\Upsilon_1(t) &\leq -N\alpha\beta^{-1}\int_0^1 \Big(\phi_x\hat{k}*\phi_x + 2\phi_{tx}\hat{k}*\phi_{tx} + \phi_{ttx}\hat{k}*\phi_{ttx} \\ &\quad + \phi_{xx}\hat{k}*\phi_{xx} + \phi_{txx}\hat{k}*\phi_{txx}\Big)dx \\ &\quad - C_3(n(t,v,\phi)+n(t,v_t,\phi_t)) + \frac{C_3}{8C_6}\mathcal{M}(t,v,\phi) \\ &\quad + C_{15}(\epsilon+\delta)\mathcal{M}(t,v,\phi) + C_4\Big(\parallel \hat{k}*\phi_x \parallel^2 + 2 \parallel \hat{k}*\phi_{tx} \parallel^2 \\ &\quad + \parallel \hat{k}*\phi_{ttx} \parallel^2 + \parallel \hat{k}*\phi_{xx} \parallel^2 + \parallel \hat{k}*\phi_{txx} \parallel^2 \Big) + \frac{a_1^2}{4} \parallel \hat{k}*\phi_{tx} \parallel^2 \\ &\quad + C_{16}\delta\Big(\parallel \hat{k}*\phi_{tx} \parallel^2 + \parallel \hat{k}*\phi_{xx} \parallel^2 + \parallel \hat{k}*\phi_{txx} \parallel^2 \Big) \\ &\quad + \Big[\lambda_1(\hat{k}'(t))^2 + \lambda_2(\hat{k}(t))^2\Big] \parallel \phi_{0xx} \parallel^2 + \lambda_3(\hat{k}'(t))^2(\parallel \phi_{1x} \parallel^2 + \parallel \phi_{0x} \parallel^2) \\ &\quad + \lambda_4\Big[(k_2(t))^2 + ((k_1*k_2)(t))^2 + ((k_1*k_2)'(t))^2\Big] \parallel p_{0x} \parallel^2, \quad (2.2.84)\end{aligned}$$

which, together with (2.2.82)–(2.2.83), yields that if $\epsilon + \delta$ is small enough,

$$\begin{aligned}\frac{d}{dt}\Upsilon(t) \le & -N\alpha\beta^{-1}\int_0^1 \Big(\phi_x\hat{k} * \phi_x + 2\phi_{tx}\hat{k} * \phi_{tx} + \phi_{ttx}\hat{k} * \phi_{ttx} \\ & + \phi_{xx}\hat{k} * \phi_{xx} + \phi_{txx}\hat{k} * \phi_{txx}\Big)dx \\ & - \frac{C_3}{2}\Big(n(t, v, \phi) + n(t, v_t, \phi_t)\Big) \\ & + \Big(2C_4 + \frac{a_1^2}{4} + C_{16}\delta + C_{14}\beta_3\Big)\Big(\| \hat{k} * \phi_x \|^2 + \| \hat{k} * \phi_{tx} \|^2 \\ & + \| \hat{k} * \phi_{ttx} \|^2 + \| \hat{k} * \phi_{xx} \|^2 + \| \hat{k} * \phi_{txx} \|^2 \Big) \\ & + \Big[\lambda_1(\hat{k}'(t))^2 + \lambda_2(\hat{k}(t))^2\Big] \| \phi_{0xx} \|^2 + \lambda_3(\hat{k}'(t))^2(\| \phi_{1x} \|^2 + \| \phi_{0x} \|^2) \\ & + \lambda_4\Big[(k_2(t))^2 + ((k_1 * k_2)(t))^2 + ((k_1 * k_2)'(t))^2\Big] \| p_{0x} \|^2 . \end{aligned} \tag{2.2.85}$$

Integrating (2.2.85) with respect to t, using Lemma 2.2.2, (2.2.79)–(2.2.82) and (2.2.59), taking δ and ϵ small enough, we deduce

$$\begin{aligned} & \Upsilon(t) + \frac{C_3}{2}\int_0^t \Big(n(\tau, v, \phi) + n(\tau, v_t, \phi_t)\Big)d\tau \\ & \quad + C_{17}\int_0^t \left[\sum_{i=0}^{2} \| \hat{k} * \partial_t^i \phi_x \|^2 + \sum_{i=0}^{1} \| \hat{k} * \partial_t^i \phi_x \|^2\right] d\tau \\ & \le (\beta_1 + \beta_2)\frac{\alpha_1\alpha_2\epsilon^2}{2} + \tilde{k}_1(0)\Big[2(\beta_1 + \beta_2) + \frac{N^2\alpha^2 C_4}{\beta_1\beta_2}\Big]\alpha_1\alpha_2\epsilon^2 + \eta_0\epsilon^2 \\ & = \alpha_3\epsilon^2. \end{aligned} \tag{2.2.86}$$

Thus it follows from (2.2.57), (2.2.60), (2.2.82) and (2.2.86) that for any $t \in [0, t_1)$,

$$\begin{aligned} & \mathcal{M}(t, v, \phi) + \frac{2C_3}{\beta_1}\int_0^t \mathcal{M}(\tau, v, \phi)d\tau \\ & \quad + \frac{4C_6C_{17}}{\beta_1}\int_0^t \left(\sum_{i=0}^{2} \| \hat{k} * \partial_t^i \phi_x \|^2 + \sum_{i=0}^{1} \| \hat{k} * \partial_t^i \phi_x \|^2\right) d\tau \\ & \le \frac{4C_6\alpha_3\epsilon^2}{\beta_1} = (\alpha_0 - \alpha_1\alpha_2)\epsilon^2. \end{aligned} \tag{2.2.87}$$

Letting $t \to t_1$ in (2.2.87), we have

$$\mathcal{M}(t_1, v, \phi) \le (\alpha_0 - \alpha_1\alpha_2)\epsilon^2 < \alpha_0\epsilon^2,$$

which contradicts the definition of t_1, By repeating the same procedure, taking ϵ even smaller if necessary, and using the continuity of $\mathcal{M}(t, v, \phi)$, (2.2.87) is established for all $t > 0$. On the other hand, it is easy to verify that for all $t > 0$,

$$C_{18}^{-1}\mathcal{M}(t,\theta,v)e^{2\delta t} \le \mathcal{M}(t,v,\phi) \le C_{18}\mathcal{M}(t,\theta,v)e^{2\delta t}. \tag{2.2.88}$$

By (2.2.18), we deduce

$$\begin{aligned}
&\frac{d}{dt}\left(\sum_{i=0}^{1} \| (\hat{k} * \partial_t^i\phi_x)(t) \|^2 + \| (\hat{k} * \partial_t^i\phi_{xx})(t) \|^2\right)\\
&\le C\Big(\sum_{i=0}^{1}(\| (\hat{k} * \partial_t^i\phi_x)(t) \|^2 + \| (\partial_t^i\phi_x)(t) \|^2)\\
&\quad + \| (\hat{k} * \partial_t^i\phi_{xx})(t) \|^2 + \| \phi_{xx}(t) \|^2\Big)\\
&\le C\Big(\sum_{i=0}^{1} \| (\hat{k} * \partial_t^i\phi_x)(t) \|^2 + \| (\hat{k} * \partial_t^i\phi_{xx})(t) \|^2 + \mathcal{M}(t,v,\phi)\Big).
\end{aligned} \tag{2.2.89}$$

Integrating (2.2.89) with respect to t, and exploiting (2.2.87), we finally obtain

$$\sum_{i=0}^{1}\Big(\| (\hat{k} * \partial_t^i\phi_x)(t) \|^2 + \| (\hat{k} * \partial_t^i\phi_{xx})(t) \|^2\Big) \le C. \tag{2.2.90}$$

Then by (2.2.11)–(2.2.14), (2.2.21), (2.2.87), (2.2.88) and (2.2.90), the proof of Theorem 2.1.1 is now complete. □

2.3 Bibliographic Comments

Thermoelastic equations describe the elastic and the thermal behavior of elastic, heat conductive media, in particular the reciprocal actions between elastic stresses and temperature differences [27, 53, 122].

The classical model of thermoelasticity, constructed on the basis of the Fourier's law, provides good approximations for the description in a wide range of engineering applications. In the simplest case of a homogeneous isotropic medium, we have the following equations in two or three space dimensions

$$\begin{aligned}
&U_{tt} - \Big((2\mu+\lambda)\nabla\nabla' - \mu\nabla\times\nabla\times\Big)U + \gamma\nabla\theta = b,\\
&\delta\theta_t - \kappa\Delta\theta + \gamma\nabla' U_t = r,
\end{aligned}$$

where $\mu, \lambda, \gamma, \delta, b, r$ and κ are constants (μ, λ are the Lamé moduli) satisfying

$$\mu > 0, \quad 2\mu + \lambda > 0,$$

and

$$\delta, \kappa > 0, \quad \gamma \neq 0.$$

Notice that in two space dimensions, the rotation of a scalar field f in $\mathbb{R}^2$ is defined to be the vector field

$$\nabla \times f := (\partial_2 f, -\partial_1 f)'$$

and the rotation of a vector field $F = (F_1, F_2)'$ in $\mathbb{R}^2$ is defined to the scalar

$$\nabla \times F := \partial_1 F_2 - \partial_2 F_1.$$

In particular, the formula

$$\Delta = \nabla\nabla' - \nabla \times \nabla$$

holds in $\mathbb{R}^2$ and $\mathbb{R}^3$.

However, this model leads to the paradox of the infinite propagation speed of heat pulse and in some practical situations may lead to an inadequate description of heat conduction. In order to eliminate these shortcomings of classical thermoelasticity, many hyperbolic thermoelastic models have been developed from the middle of the last century. Recently, Green and Naghdi [89, 90] re-examined the classical Fourier's law in thermoelasticity, instead of the classical entropy inequality, used a general entropy balance and, introducing a new thermal variable, proposed three models, based on the different material responses, labeled as types I, II and III. The linearized version of the first model leads to the Fourier law, the linearized version of both types II and III models whose constitutive assumptions on the heat flux vector are different from the Fourier's law allows heat transmission at a finite speed.

Let us now recall the classical model for linear thermoelastic systems of types I, II and III which take the following forms, respectively.
Classical model of type I:

$$\begin{cases} U_{tt} - \Big((2\mu + \lambda)\nabla\nabla' - \mu\nabla \times \nabla \times\Big)U + \gamma\nabla\theta = 0, \\ \delta\theta_t - \kappa\Delta\theta + \gamma\nabla \cdot U_t = 0. \end{cases} \tag{2.3.1}$$

Model of type II:

$$\begin{cases} U_{tt} - \Big((2\mu + \lambda)\nabla\nabla' - \mu\nabla \times \nabla \times\Big)U + \gamma\nabla\theta = 0, \\ \delta\theta_{tt} - \kappa\Delta\theta + \gamma\nabla \cdot U_{tt} = 0. \end{cases} \tag{2.3.2}$$

Model of type III:

$$\begin{cases} U_{tt} - \Big((2\mu + \lambda)\nabla\nabla' - \mu\nabla \times \nabla \times\Big)U + \gamma\nabla\theta = 0, \\ \delta\theta_{tt} - \kappa\Delta\theta - \kappa\Delta\theta_t + \gamma\nabla \cdot U_{tt} = 0. \end{cases} \tag{2.3.3}$$

In the above, $U = U(x, t)$ is the displacement vector, $\theta = \theta(x, t)$ denotes the temperature.

Moreover, the three-dimensional thermoelastic equations with second sound obeying Cattaneo's law takes the form

$$\begin{cases} U_{tt} - \mu\Delta U - (\mu + \lambda)\nabla divU + \beta\nabla\theta = 0, \\ \theta_t + \gamma divq + \delta divU_t = 0, \\ \tau_0 q_t + q + \kappa\nabla\theta = 0, \end{cases} \tag{2.3.4}$$

here, $\mu, \beta, \gamma, \delta$ and κ are positive constants, U and q are two unknown vector functions, while θ is a unknown scalar function. Constant $\tau_0 > 0$ is the so-called relaxation parameter.

For the thermoelasticity of type I, there are many works (see, e.g., [47, 110, 120, 154, 183, 184, 185, 186, 274, 106, 122, 155, 240]) on the existence, uniqueness and asymptotic behavior of solutions of the linear system. Slemrod [257] proved the global existence, uniqueness and asymptotic stability of classical smooth solutions; Shibata [255] considered the initial boundary value problem with the boundary conditions $u_x - \gamma\theta = 0, \theta_x = 0 \ \ (x = 0, l)$; Racke, Shibata and Zheng [240] obtained the global existence and uniqueness of solutions for the nonlinear thermoelastic system of type I with small initial data; Jiang [120] proved an exponential decay result for solutions of the equations of linear, homogeneous, isotropic thermoelasticity in bounded regions in two or three space dimensions; Racke [234] considered the Cauchy problem in three-dimensional nonlinear thermoelasticity for a medium which is homogeneous and initially isotropic; Lebeau and Zuazua [143], by a decoupling method, reduced the problem to an observability inequality for the Lamé system in linear elasticity and more precisely to whether the total energy of solutions can be estimated in terms of the energy concentrated on its longitudinal component, and showed that when the domain is convex, the decay rate is never uniform, and, in three space dimensions, the lack of uniform decay may be due to a critical polarization of the energy on the transversal component of the displacement; Muñoz Rivera and Qin [186] proved the global existence, uniqueness, and asymptotic behavior of solutions for $1D$ nonlinear thermoelasticity with thermal memory subject to Dirichlet-Dirichlet boundary conditions; Muñoz Rivera and Qin [189] proved the global existence, uniqueness and asymptotic behavior of solutions for the one-dimensional nonlinear thermoelasticity with thermal memory subject to Dirichlet-Dirichlet boundary conditions. When Cattaneo's law substitutes Fourier's law, results concerning existence, blow-up, and asymptotic behavior of smooth (weak) solutions have been established by several authors (see, e.g., [166, 167, 169, 235, 236, 266, 275, 281]).

Now we recall some results on the thermoelastic systems with second sound. Tarabek [266] treated problems related to system (2.1.1)–(2.1.3) in both bounded and unbounded domains and established global existence results for small initial data, and showed that these classical solutions tend to their equilibria as t tends to infinity. Concerning the asymptotic behavior, Racke [235] discussed the global existence and decay exponentially to the equilibrium state for one-dimensional case. For the multi-dimensional case ($n = 2, 3$), Racke [236] established an existence result for homogeneous linear problem; Messaoudi [164] investigated the situation where a nonlinear source term is competing with a damping caused by the heat conduction and established a local existence result, and further prove that solutions with negative energy blow up in a finite time. Later on, Messaoudi and Said-Houari [169] proved that the exponential stability in one-dimensional nonlinear thermoelasticity with second sound. Their work generalized earlier ones in [165, 166, 167] to thermoelasticity with second sound. Racke and Wang [239] proved asymptotic behavior of discontinuous solutions to thermoelastic systems with second sound. The Cauchy problem of the linear thermoelastic system with second sound was also studied by Wang and Wang [275], and Yang and Wang [281]. In particular, if $k_1 = 0,\ q = -k\theta_x$, then the system (2.1.1)–(2.1.3) reduces to the system of a classical thermoelasticity, in which the heat flux is given by Fourier's law instead of Cattaneo's law and results concerning existence, blow-up, and asymptotic behavior of smooth (weak) solutions have been established by several authors over the past two decades (see, e.g., [47, 89, 90, 120, 169, 183, 210, 226, 257, 274]); if $k_1 \neq 0, q = -k_1 * \theta_x$, Muñoz Rivera and Qin [189] studied the global existence, uniqueness, and asymptotic behavior of solutions to the equations of thermoelastic system subject to Dirichlet-Dirichlet boundary conditions. When the heat flux obeys the theory of Gurtin and Pipkin, that is, $q = -k_1 * \theta_x$, Fatori and Muñoz Rivera [74] established the energy decay for a linear hyperbolic thermoelastic system provided the relaxation kernel $k_1(t)$ is a strongly positive definite and decays exponentially. For the models of linear and nonlinear thermoelastic plates, Bucci and Chueshov [24] proved the existence of a compact, finite dimensional, global attractor for a coupled PDE system comprising a nonlinearly damped semilinear wave equation and a nonlinear system of thermoelastic plate equations, without any mechanical (viscous or structural) dissipation in the plate component, a major part in the proof is played by an estimate-known as stabilizability estimate-which shows that the difference of any two trajectories can be exponentially stabilized to zero, modulo a compact perturbation. Chueshov and Lasiecke [40] studied asymptotic behavior of solutions corresponding to von Karman thermoelastic plates, a distinct feature of this work is that the model considered has no added dissipation-particularly mechanical dissipation typically added to plate equation when long time-behavior is considered, thus, the model consists of undamped oscillatory plate equation strongly coupled with heat equation, nevertheless the authors were able to showed that the ultimate (asymptotic) behavior of the von Karman evolution is described by finite dimensional global attractor. In addition, the obtained estimate for the dimension and the size of the attractor are independent of the rotational inertia parameter γ and heat/thermal capacity k, where the former was known to change the character of dynamics from hyperbolic ($\gamma > 0$) to parabolic like ($\gamma = 0$). Meyvaci [180] studied the problem of continuous dependence on parameters of strong solutions of the initial boundary value problem for the linear thermoelastic plate equation.

For the 2D Timoshenko systems with second sound, we notice the works of Fastovska [72] and Grobbelaar-Van Dalsen. In [94], the authors are concerned with the strong stabilization of models for the Reissner-Mindlin plate equations with second sound, that is, models that include thermal effects described according to Cattaneo's law of heat conduction instead of Fourier's law in classical thermoelasticity. Two models will be considered which are distinct with respect to the property of compactness or non-compactness of the resolvent of the generator of the underlying semigroup. In accordance with the compactness or non-compactness of the resolvent operator, a different criterion for strong stability is implemented to achieve the strong stabilization of each model. Recently, Clark et al. [43] gave a mathematical treatment of a model for small vertical vibrations of an elastic membrane coupled with a heat equation and subject to nonlinear mixed boundary conditions. They established the existence, uniqueness, and a uniform decay rate for global solutions to this nonlinear nonlocal thermoelastic coupled system under nonlinear boundary conditions. Later on, Clark and Guardia [42] dealt with the global existence and uniqueness of solutions, and uniform stabilization of the energy of an initial-boundary value problem for a thermoelastic system with nonlinear terms of nonlocal kind. The asymptotic stabilization of the energy of system is obtained without any dissipation mechanism acting in the displacement variable u of the membrane equation. Fatori et al. [73] considered the long-time behavior of a class of thermoelastic plates with nonlinear strain, they established the existence of global and exponential attractors for the strongly damped problem through a stabilizability inequality. In addition, for the weakly damped problem, they established the exponential stability of its Galerkin semiflows.

We would like to mention other works in [2, 26, 41, 44, 139, 145, 138, 188, 272] for related models.

Chapter 3
Energy Decay for a Timoshenko-Type System with a Past History

3.1 Introduction

In this chapter, we shall consider the following Timoshenko-type system

$$\begin{cases} \rho_1\varphi_{tt} - k(\varphi_x + \psi)_x = 0, \\ \rho_2\psi_{tt} - b\psi_{xx} + \int_0^{+\infty} g(s)\psi_{xx}(t-s)ds + k(\varphi_x + \psi) + \delta\theta_x = 0, \\ \rho_3\theta_t - \beta\theta_{xx} + \delta\psi_{xt} = 0 \end{cases} \tag{3.1.1}$$

with positive constants $\rho_1, \rho_2, \rho_3, k, b, \beta, \delta$ together with initial conditions

$$\varphi(\cdot, 0) = \varphi_0,\ \varphi_t(\cdot, 0) = \varphi_1,\ \psi(\cdot, 0) = \psi_0,\ \psi_t(\cdot, 0) = \psi_1,\ \theta(\cdot, 0) = \theta_0 \tag{3.1.2}$$

and boundary conditions

$$\varphi(0, t) = \varphi(1, t) = \psi(0, t) = \psi(1, t) = \theta(0, t) = \theta(1, t) = 0, \tag{3.1.3}$$

where the functions φ, ψ and θ depend on $(x, t) \in [0, 1] \times [0, +\infty)$ and model the transverse displacement of a beam with reference configuration $(0, 1) \subset \mathbb{R}$, the rotation angle of a filament and the temperature difference respectively.

In this chapter, we have extended and improved those results in [81] with weaker conditions on g, that is, we shall use a different method (the multiplier techniques, which was also used in [174]) from that in [81] to prove the exponential stability result which was also obtained in [81] only for the case of equal wave speeds. Furthermore, we also prove a polynomial stability result which has been never studied before. We adopt the results in this chapter from [219].

Y. Qin and Z. Ma, *Global Well-posedness and Asymptotic Behavior of the Solutions to Non-classical Thermo(visco)elastic Models*,
DOI 10.1007/978-981-10-1714-8_3

3.2 Preliminaries

In order to state our main result, we make the following hypotheses:
(H1). $g : \mathbb{R}^+ \to \mathbb{R}^+$ is a differentiable function such that

$$g(0) > 0, \quad b - \int_0^{+\infty} g(s)ds = \bar{b} > 0.$$

(H2). There exist constants $\gamma > 0$ and $1 \le p < 3/2$ such that

$$g'(s) \le -\gamma g^p(s), \quad \forall s \ge 0.$$

Following the idea of [48], we introduce

$$\eta^t(x,s) = \psi(x,t) - \psi(x,t-s), \quad s \ge 0, \tag{3.2.1}$$

and obtain the following initial and boundary conditions

$$\eta^t(x,0) = 0, \quad \eta^t(0,s) = \eta^t(1,s) = 0, \quad \eta^0(x,s) = \eta_0(x,s), \quad for \ \ all \ s,t \ge 0. \tag{3.2.2}$$

Clearly, (3.2.1) gives

$$\eta_t^t(x,s) + \eta_s^t(x,s) = \psi_t(x,t). \tag{3.2.3}$$

Remark 3.2.1 Under hypotheses (H1)-(H2), it is easy to verify that

$$G_0 = \int_0^{+\infty} g^{1/2}(s)ds < +\infty, \quad G_p = \int_0^{+\infty} g^{2-p}(s)ds < +\infty, \quad 1 \le p < 3/2. \tag{3.2.4}$$

From the semigroup theory (see, e.g., [155, 287]), we have the following existence and regularity result, for the explicit proofs, we refer the reader to [81].

Lemma 3.2.1 *Let $(\varphi_0, \varphi_1), (\psi_0, \psi_1) \in H_0^1(0,1) \times L^2(0,1), \theta_0 \in L^2(0,1)$. Assume that (H1) is satisfied, then problem (3.1.1) has a unique global weak solution $(\varphi(t), \psi(t), \theta(t))$ verifying*

$$\varphi(t), \psi(t) \in C(\mathbb{R}^+, H_0^1(0,1)) \cap C^1(\mathbb{R}^+, L^2(0,1)), \quad \theta(t) \in C(\mathbb{R}^+, L^2(0,1)).$$

Moreover, if

$$(\varphi_0, \varphi_1), (\psi_0, \psi_1) \in (H_0^2(0,1) \times H_0^1(0,1)) \times H_0^1(0,1), \quad \theta_0 \in H_0^1(0,1),$$

then the solution $(\varphi(t), \psi(t), \theta(t))$ *satisfies*

$$\begin{cases} \varphi(t), \psi(t) \in C(\mathbb{R}^+, H_0^2(0,1) \times H_0^1(0,1)) \cap C^1(\mathbb{R}^+, H_0^1(0,1)) \cap C^2(\mathbb{R}^+, L^2(0,1)), \\ \theta(t) \in C(\mathbb{R}^+, H_0^2(0,1) \times H_0^1(0,1)) \cap C^1(\mathbb{R}^+, H_0^1(0,1)). \end{cases}$$

3.3 Case of Equal Wave-Speeds

In this section, we shall state and prove a decay result in the case of equal wave speeds propagation. First, combining (3.1.1)–(3.1.3) and (3.2.1)–(3.2.2), we give a reformulation system as

$$\begin{cases} \rho_1\varphi_{tt} - k(\varphi_x + \psi)_x = 0, \\ \rho_2\psi_{tt} - \bar{b}\psi_{xx} - \int_0^{+\infty} g(s)\eta_{xx}^t(t-s)ds + k(\varphi_x + \psi) + \delta\theta_x = 0, \\ \rho_3\theta_t - \beta\theta_{xx} + \delta\psi_{xt} = 0, \\ \eta_t^t(x,s) + \eta_s^t(x,s) - \psi_t(x,t) = 0 \end{cases} \tag{3.3.1}$$

in $(0, 1) \times (0, +\infty)$ together with initial conditions and boundary conditions

$$\varphi(\cdot,0) = \varphi_0,\ \varphi_t(\cdot,0) = \varphi_1,\ \psi(\cdot,0) = \psi_0,\ \psi_t(\cdot,0) = \psi_1,\ \theta(\cdot,0) = \theta_0,\ \eta^t(x,0) = 0, \tag{3.3.2}$$

$$\varphi(0,t) = \varphi(1,t) = \psi(0,t) = \psi(1,t) = \theta(0,t) = \theta(1,t) = \eta^t(0,s) = \eta^t(1,s) = 0. \tag{3.3.3}$$

The associated energy of system (3.3.1) is given by

$$\begin{aligned} E(t) &= \frac{1}{2}\int_0^1 \left(\rho_1\varphi_t^2 + \rho_2\psi_t^2 + \rho_3\theta^2 + k \mid \varphi_x + \psi \mid^2 + \bar{b}\psi_x^2\right) dx \\ &+ \frac{1}{2}\int_0^1 \int_0^{+\infty} g(s) \mid \eta_x^t(x,s) \mid^2 dsdx. \end{aligned} \tag{3.3.4}$$

Let $L_g^2(\mathbb{R}^+, H_0^1)$ be the Hilbert space of H_0^1-valued functions on $\mathbb{R}^+$, endowed with the norm

$$\| w \|_{L_g^2(\mathbb{R}^+, H_0^1)} = \left(\int_0^1 \int_0^{+\infty} g(s) \mid w_x(x,s) \mid^2 dsdx\right)^{1/2}.$$

We are now ready to state our main stability result.

Theorem 3.3.1 *Let* $(\varphi_0, \varphi_1), (\psi_0, \psi_1) \in H_0^1(0,1) \times L^2(0,1), \eta_0^t \in L_g^2\Big(\mathbb{R}^+, H_0^1(0,1)\Big)$ *and* $\theta_0 \in L^2(0,1)$. *Assume that (H1)-(H2) and* $\frac{\rho_1}{k} = \frac{\rho_2}{b}$ *are satisfied, then there exist two positive constants* C *and* ω, *such that*

$$E(t) \le Ce^{-\omega t}, \qquad p = 1, \quad for \ \ all \ t \ge 0, \tag{3.3.5}$$

$$E(t) \le \frac{C}{(t+1)^{1/(p-1)}}, \quad p > 1, \quad for \ \ all \ t \ge 0. \tag{3.3.6}$$

The proof of our result will be established through the following several lemmas.

Lemma 3.3.1 *Let $(\varphi, \psi, \theta, \eta^t)$ be the solution of problem (3.3.1), then we have*

$$E'(t) = -\beta \int_0^1 \theta_x^2 dx + \frac{1}{2}\int_0^1 \int_0^{+\infty} g'(s) \mid \eta_x^t(x,s) \mid^2 dsdx \le 0. \tag{3.3.7}$$

Proof Multiplying (3.3.1)–(3.3.3) by φ_t, ψ_t and θ respectively and integrating the result over (0, 1), using hypotheses (H1)-(H2) and summing up, we can obtain (3.3.7). □

Lemma 3.3.2 *Let $(\varphi, \psi, \theta, \eta^t)$ be the solution of problem (3.3.1), then we have for $1 < p < 3/2$,*

$$\left(\int_0^1 \int_0^{+\infty} g(s) \mid \eta_x^t(x,s) \mid^2 dsdx\right)^{2p-1} \le C \int_0^1 \int_0^{+\infty} g^p(s) \mid \eta_x^t(x,s) \mid^2 dsdx \tag{3.3.8}$$

for a constant $C > 0$.

Proof Using Hölder's inequality, it is straightforward to see that, for any $r > 1$, we have

$$\int_0^1 \int_0^{+\infty} g(s) \mid \eta_x^t(x,s) \mid^2 dsdx = \int_0^1 \int_0^{+\infty} g^{\frac{1}{2r}}(s) \mid \eta_x^t(x,s) \mid^{\frac{2r-2}{r}} dsdx$$
$$\le \left(\int_0^1 \int_0^{+\infty} g^{\frac{1}{2}}(s) \mid \eta_x^t(x,s) \mid^2 dsdx\right)^{\frac{1}{r}} \left(\int_0^1 \int_0^{+\infty} g^{\frac{2r-1}{2r-2}}(s) \mid \eta_x^t(x,s) \mid^2 dsdx\right)^{\frac{r-1}{r}}.$$

Then from (3.2.1), (3.2.4), (3.3.4) and (3.3.7) it follows

$$\int_0^1 \int_0^{+\infty} g^{\frac{1}{2}}(s) \mid \eta_x^t(x,s) \mid^2 dsdx \le 2E(0) \int_0^{+\infty} g^{1/2}(s)ds = 2G_0E(0).$$

Taking $r = (2p-1)/(2p-2)$, we can derive (3.3.8). □

Lemma 3.3.3 *Let $(\varphi, \psi, \theta, \eta^t)$ be the solution of problem (3.3.1)–(3.3.3), then we have for $1 \le p \le 3/2$,*

$$\int_0^1 \left(\int_0^{+\infty} g(s)\eta_x^t(x,s)ds\right)^2 dx \le G_p \int_0^1 \int_0^{+\infty} g^p(s) \mid \eta_x^t(x,s) \mid^2 dsdx. \tag{3.3.9}$$

Proof Using the Cauchy–Schwarz's inequality, we get

$$\int_0^{+\infty} g(s)\eta_x^t(x,s)ds = \int_0^{+\infty} g^{1-p/2}(s)g^{p/2}(s)\eta_x^t(x,s)ds \le \left(\int_0^{+\infty} g^{2-p}(s)ds\right)^{1/2} \times \left(\int_0^{+\infty} g^p(s) \mid \eta_x^t(x,s) \mid^2 ds\right)^{1/2}.$$

Therefore (3.3.9) follows from (3.2.4). □

Now we are going to construct a Lyapunov functional $\mathcal{F}(t)$ equivalent to $E(t)$. To this end, we define several functionals which allow us to obtain the needed estimates. Let

$$I_1 := \int_0^1 (\rho_1\varphi_t w + \rho_2\psi_t\psi)dx, \tag{3.3.10}$$

where w is the solution of

$$-w_{xx} = \psi_x, \quad w_0 = w_1 = 0.$$

Lemma 3.3.4 *Let $(\varphi, \psi, \theta, \eta^t)$ be the solution of problem (3.3.1)–(3.3.3), then we have for any $\varepsilon_1, c_1 > 0$,*

$$\frac{dI_1(t)}{dt} \le (-\bar{b} + 2c_1)\int_0^1 \psi_x^2 dx + \varepsilon_1\rho_1\int_0^1 \varphi_t^2 dx + \left(\rho_2 + \frac{\rho_1}{4\varepsilon_1}\right)\int_0^1 \psi_t^2 dx + \frac{1}{4c_1}\int_0^1 \theta_x^2 dx + \frac{G_p}{4c_1}\int_0^1\int_0^{+\infty} g^p(s) \mid \eta_x^t(x,s) \mid^2 dsdx. \tag{3.3.11}$$

Proof Taking a derivative in t of (3.3.10) and using (3.3.1), we obtain

$$\frac{dI_1(t)}{dt} = -\bar{b}\int_0^1 \psi_x^2 dx + \rho_2\int_0^1 \psi_t^2 dx - k\int_0^1 \psi^2 dx + k\int_0^1 w_x^2 dx + \rho_1\int_0^1 \rho_t w_t dx - \int_0^1 \psi_x \int_0^{+\infty} g(s)\eta_x^t(x,s)dsdx.$$

Using Young's inequality and the following inequalities

$$\int_0^1 w_x^2 dx \le C\int_0^1 \psi^2 dx \le C\int_0^1 \psi_x^2 dx,$$

$$\int_0^1 w_t^2 dx \le C\int_0^1 w_{tx}^2 dx \le C\int_0^1 \psi_t^2 dx,$$

we find that

$$\frac{dI_1(t)}{dt} \leq (-\bar{b}+c_1)\int_0^1 \psi_x^2 dx + \varepsilon_1\rho_1\int_0^1 \varphi_t^2 dx + \left(\rho_2+\frac{\rho_1}{4\varepsilon_1}\right)\int_0^1 \psi_t^2 dx + \frac{1}{4c_1}\int_0^1 \theta_x^2 dx - \int_0^1 \psi_x \int_0^{+\infty} g(s)\eta_x^t(x,s)dsdx. \tag{3.3.12}$$

By using (3.3.9), the last term on the right-hand side of (3.3.12) can be estimated as follows

$$\left| - \int_0^1 \psi_x \int_0^{\infty} g(s)\eta_x^t(x,s)dsdx\right| \leq c_1\int_0^1 \psi_x^2 dx + \frac{G_p}{4c_1}\int_0^1\int_0^{+\infty} g^p(s) \mid \eta_x^t(x,s) \mid^2 dsdx. \tag{3.3.13}$$

Inserting (3.3.13) into (3.3.12), we can obtain the desired result.

Lemma 3.3.5 *Let $(\varphi, \psi, \theta, \eta^t)$ be the solution of problem (3.3.1)–(3.3.3), then the functional $I_2(t)$ defined by*

$$I_2(t) = -\rho_2\int_0^1 \psi_t(x,t)\int_0^{+\infty} g(s)\eta^t(x,s)dsdx$$

satisfies the estimate

$$\frac{dI_2(t)}{dt} \leq -\frac{\rho_2}{g_0}\int_0^1 \psi_t^2 dx + \varepsilon_2\bar{b}^2\int_0^1 \psi_x^2 dx + \varepsilon_2 k^2\int_0^1 (\varphi_x+\psi)^2 dx + \varepsilon_2\delta^2\int_0^1 \theta_x^2 dx + G_p\left(1+\frac{3}{4\varepsilon_2}\right)\int_0^1\int_0^{+\infty} g^p(s) \mid \eta_x^t(x,s) \mid^2 dsdx - \frac{g(0)}{2\rho_2}\int_0^1\int_0^{+\infty} g'(s) \mid \eta_x^t(x,s) \mid^2 dsdx, \tag{3.3.14}$$

for any $\varepsilon_2 > 0$, where $g_0 = \int_0^{+\infty} g(s)ds$.

Proof Using (3.3.1) and Young's inequality, similarly to (3.3.11), we can easily obtain the result. □

Next we introduce the functional

$$I_3(t) := \rho_2\int_0^1 \psi_t(\varphi_x+\psi)dx + \frac{\rho_1\bar{b}}{k}\int_0^1 \psi_x\varphi_t dx + \frac{\rho_1}{k}\int_0^1 \varphi_t\int_0^{+\infty} g(s)\eta_x^t(x,s)dsdx. \tag{3.3.15}$$

Lemma 3.3.6 *Let $(\varphi, \psi, \theta, \eta^t)$ be the solution of problem (3.3.1)–(3.3.3), then for any $\varepsilon_3 > 0$, we conclude*

$$\frac{dI_3(t)}{dt} \leq \left(\varphi_x(\bar{b}\psi_x + \int_0^{+\infty} g(s)\eta_x^t(x,s)ds)\right)\Big|_{x=0}^{x=1} - k\int_0^1 (\varphi_x+\psi)^2 dx + \rho_2\int_0^1 \psi_t^2 dx$$
$$+ 2\varepsilon_3 \int_0^1 \varphi_t^2 dx + \frac{1}{4\varepsilon_3}\int_0^1 \theta_x^2 dx - g(0)C(\varepsilon_3)\int_0^1\int_0^{+\infty} g'(s) \mid \eta_x^t(x,s) \mid^2 dsdx. \tag{3.3.16}$$

Proof Differentiating $I_3(t)$ and using (3.3.1), we obtain

$$\frac{dI_3(t)}{dt} = \int_0^1 (\varphi_x+\psi)\left(\bar{b}\psi_{xx} + \int_0^{+\infty} g(s)\eta_{xx}^t(x,t-s)ds - k(\varphi_x+\psi) - \delta\theta_x\right)dx$$
$$+ \rho_2\int_0^1 \psi_t^2 dx + \bar{b}\int_0^1 \psi_x(\varphi_x+\psi)_x dx + \left(\frac{\rho_1 b}{k} - \rho_2\right)\int_0^1 \psi_{tx}\varphi_t dx$$
$$- \frac{\rho_1}{k}\int_0^1 \varphi_t \int_0^{+\infty} g(s)\eta_{sx}^t(x,s)dsdx + \int_0^1 (\varphi_x+\psi)_x \int_0^{+\infty} g(s)\eta_x^t(x,s)dsdx.$$

Using $\frac{k}{\rho_1} = \frac{b}{\rho_2}$ and Young's inequality, we conclude (3.3.16), □

We now define the following function to handle the boundary terms appearing in (3.3.16)

$$q(x) = 2 - 4x, \quad x \in [0,1].$$

Consequently, we have the following result.

Lemma 3.3.7 *Let $(\varphi, \psi, \theta, \eta^t)$ be the solution of problem (3.3.1)–(3.3.3), then we have*

$$\left(\varphi_x\left(\bar{b}\psi_x + \int_0^{+\infty} g(s)\eta_x^t(x,s)\right)\right)\Big|_{x=0}^{x=1}$$
$$\leq -\frac{\varepsilon_3}{k}\frac{d}{dt}\int_0^1 \rho_1 q(x)\varphi_t\varphi_x dx + 3\varepsilon_3\int_0^1 \varphi_x^2 dx + \frac{2\rho_1\varepsilon_3}{k}\int_0^1 \varphi_t^2 dx$$
$$- \frac{1}{4\varepsilon_3}\frac{d}{dt}\int_0^1 \rho_2 q(x)\psi_t\left(\bar{b}\psi_x + \int_0^{+\infty} g(s)\eta_x^t(x,s)ds\right)dx + 4\delta^2 c_3\int_0^1 \theta_x^2 dx$$
$$+ \frac{1}{\varepsilon_3}\left(\bar{b}^2 + \frac{\bar{b}^2}{4c_3} + \varepsilon_3^2\right)\int_0^1 \psi_x^2 dx + \frac{G_0}{4\varepsilon_3}\left(4 + \frac{1}{c_3}\right)\int_0^1\int_0^{+\infty} g^p(s) \mid \eta_x^t(x,s) \mid^2 dsdx$$
$$+ \frac{\rho_2}{4\varepsilon_3}(2\bar{b} + \varepsilon_3)\int_0^1 \psi_t^2 dx + \frac{c_3 k^2}{\varepsilon_3}\int_0^1 (\varphi_x+\psi)^2 dx$$
$$- \rho_2 g(0)c(\varepsilon_3)\int_0^1\int_0^{+\infty} g'(s) \mid \eta_x^t(x,s) \mid^2 dsdx, \tag{3.3.17}$$

for any $c_3,\ \varepsilon_3 > 0$.

Proof Using Young's inequality, we easily see that for $\varepsilon_3 > 0$,

$$\left(\varphi_x\left(\bar{b}\psi_x + \int_0^{+\infty} g(s)\eta_x^t(x,s)\right)\right)\Big|_{x=0}^{x=1}$$
$$\leq \varepsilon_3\left(\varphi_x^2(1,t) + \varphi_x^2(0,t)\right) + \frac{1}{4\varepsilon_3}\left(\bar{b}\psi_x(0,t) + \int_0^{+\infty} g(s)\eta_x^t(0,s)ds\right)^2$$

$$+\frac{1}{4\varepsilon_3}\left(\bar{b}\psi_x(1,t)+\int_0^{+\infty}g(s)\eta_x^t(1,s)ds\right)^2. \tag{3.3.18}$$

Exploiting

$$\begin{aligned}&\frac{d}{dt}\int_0^1 \rho_2 q(x)\psi_t\left(\bar{b}\psi_x+\int_0^{+\infty}g(s)\eta_x^t(x,s)ds\right)dx\\&=\int_0^1 \rho_2 q(x)\psi_{tt}\left(\bar{b}\psi_x+\int_0^{+\infty}g(s)\eta_x^t(x,s)ds\right)dx\\&\quad+\int_0^1 \rho_2 q(x)\psi_t\left(\bar{b}\psi_{tx}+\int_0^{+\infty}g(s)\eta_{tx}^t(x,s)ds\right)dx\end{aligned}$$

and (3.3.1), we get

$$\begin{aligned}&\frac{d}{dt}\int_0^1 \rho_2 q(x)\psi_t\left(\bar{b}\psi_x+\int_0^{+\infty}g(s)\eta_x^t(x,s)ds\right)dx\\&=\int_0^1 q(x)\left(\bar{b}\psi_{xx}+\int_0^{+\infty}g(s)\eta_{xx}^t(x,s)ds-k(\varphi_x+\psi)-\delta\theta_x\right)\\&\quad\times\left(\bar{b}\psi_x+\int_0^{+\infty}g(s)\eta_x^t(x,s)ds\right)dx\\&\quad+\int_0^1 \rho_2 q(x)\psi_t\left(\bar{b}\psi_x+\int_0^{+\infty}g(s)\eta_x^t(x,s)ds\right)dx.\end{aligned} \tag{3.3.19}$$

A direct calculation shows

$$\begin{aligned}&\int_0^1 q(x)\left(\bar{b}\psi_{xx}+\int_0^{+\infty}g(s)\eta_{xx}^t(x,s)ds\right)\left(\bar{b}\psi_x+\int_0^{+\infty}g(s)\eta_x^t(x,s)ds\right)dx\\&=-\frac{1}{2}\int_0^1 q'(x)\left(\bar{b}\psi_x+\int_0^{+\infty}g(s)\eta_x^t(x,s)ds\right)^2dx\\&\quad+\left(\frac{q(x)}{2}\left(\bar{b}\psi_x+\int_0^{+\infty}g(s)\eta_x^t(x,s)ds\right)^2\right)\Big|_{x=0}^{x=1}.\end{aligned} \tag{3.3.20}$$

By using (3.3.1), the last term in (3.3.19) can be treated as follows

$$\begin{aligned}&\int_0^1 \rho_2 q(x)\psi_t\left(\bar{b}\psi_x+\int_0^{+\infty}g(s)\eta_x^t(x,s)ds\right)dx\\&=\rho_2\bar{b}\int_0^1 q(x)\psi_t\psi_{tx}dx+\rho_2\int_0^1 q(x)\psi_t\int_0^{+\infty}g(s)\eta_{tx}^t(x,s)dsdx\\&=-\frac{\rho_2\bar{b}}{2}\int_0^1 q'(x)\psi_t^2dx+\rho_2\int_0^1 q(x)\psi_t\int_0^{+\infty}g(s)\eta_{tx}^t(x,s)dsdx\\&=-\frac{\rho_2\bar{b}}{2}\int_0^1 q'(x)\psi_t^2dx+\rho_2\int_0^1 q(x)\psi_t\int_0^{+\infty}g(s)(\psi_{tx}(t)-\eta_{sx}^t(x,s))dsdx\end{aligned}$$

$$= -\frac{\rho_2 \bar{b}}{2} \int_0^1 q'(x)\psi_t^2 dx + g_0\rho_2 \int_0^1 q(x)\psi_t\psi_{tx} dx - \rho_2 \int_0^1 q(x)\psi_t \int_0^{+\infty} g(s)\eta_{sx}^t(x,s) ds dx$$

$$= -\frac{\rho_2(\bar{b}+g_0)}{2} \int_0^1 q'(x)\psi_t^2 dx + \rho_2 \int_0^1 q(x)\psi_t \int_0^{+\infty} g'(s)\eta_x^t(x,s) ds dx. \quad (3.3.21)$$

Inserting (3.3.20) and (3.3.21) into (3.3.19), we arrive at

$$\begin{aligned}
&\left(\bar{b}\psi_x(0,t) + \int_0^{+\infty} g(s)\eta_x^t(0,s)ds\right)^2 + \left(\bar{b}\psi_x(1,t) + \int_0^{+\infty} g(s)\eta_x^t(1,s)ds\right)^2 \\
&= -\frac{d}{dt}\int_0^1 \rho_2 q(x)\psi_t \left(\bar{b}\psi_x + \int_0^{+\infty} g(s)\eta_x^t(x,s)ds\right) dx \\
&\quad + 2\int_0^1 \left(\bar{b}\psi_x + \int_0^{+\infty} g(s)\eta_x^t(x,s)ds\right)^2 dx \\
&\quad - k\int_0^1 q(x)(\varphi_x+\psi)\left(\bar{b}\psi_x + \int_0^{+\infty} g(s)\eta_x^t(x,s)ds\right) dx - \delta \int_0^1 q(x)\theta_x \\
&\left(\bar{b}\psi_x + \int_0^{+\infty} g(s)\eta_x^t(x,s)ds\right) dx \\
&\quad + 2\rho_2(\bar{b}+g_0)\int_0^1 \psi_t^2 dx + \rho_2\int_0^1 q(x)\psi_t \int_0^{+\infty} g'(s)\eta_x^t(x,s) ds dx. \quad (3.3.22)
\end{aligned}$$

We now estimate each term on the right-hand side of (3.3.22), by exploiting Young's inequality and (3.3.9) as follows: The second, third and fourth terms can be estimated as

$$\begin{aligned}
2\int_0^1 \left(\bar{b}\psi_x + \int_0^{+\infty} g(s)\eta_x^t(x,s)ds\right)^2 dx &\le 4\bar{b}^2\int_0^1 \psi_x^2 dx \\
&\quad + 4G_p \int_0^1\int_0^{+\infty} g^p(s) \mid \eta_x^t(x,s) \mid^2 ds dx,
\end{aligned} \quad (3.3.23)$$

$$\begin{aligned}
&\left|k\int_0^1 q(x)(\varphi_x+\psi)\left(\bar{b}\psi_x + \int_0^{+\infty} g(s)\eta_x^t(x,s)ds\right)dx\right| \\
&\le 2k\left|\int_0^1 (\varphi_x+\psi)\left(\bar{b}\psi_x + \int_0^{+\infty} g(s)\eta_x^t(x,s)ds\right)dx\right| \\
&\le 4k^2c_3\int_0^1 (\varphi_x+\psi)^2 dx + \frac{1}{4c_3}\int_0^1 \left(\bar{b}\psi_x + \int_0^{+\infty} g(s)\eta_x^t(x,s)ds\right)^2 dx \\
&\le 4k^2c_3\int_0^1 (\varphi_x+\psi)^2 dx + \frac{\bar{b}^2}{2c_3}\int_0^1 \psi_x^2 dx + \frac{G_p}{2c_3}\int_0^1\int_0^{+\infty} g^p(s) \mid \eta_x^t(x,s) \mid^2 ds dx, \quad (3.3.24)
\end{aligned}$$

and

$$\begin{aligned}
&\Big|\delta\int_0^1 q(x)\theta_x\Big(\bar b\psi_x+\int_0^{+\infty}g(s)\eta_x^t(x,s)ds\Big)dx\Big|\\
&\le 2\delta\Big|\int_0^1 \theta_x\Big(\bar b\psi_x+\int_0^{+\infty}g(s)\eta_x^t(x,s)ds\Big)dx\Big|\\
&\le 4\delta^2 c_3\int_0^1 \theta_x^2\frac{1}{4c_3}\int_0^1\Big(\bar b\psi_x+\int_0^{+\infty}g(s)\eta_x^t(x,s)ds\Big)^2dx\\
&\le 4k^2c_3\int_0^1\theta_x^2dx+\frac{\bar b^2}{2c_3}\int_0^1\psi_x^2dx+\frac{G_p}{2c_3}\int_0^1\int_0^{+\infty}g^p(s)\mid\eta_x^t(x,s)\mid^2dsdx.
\end{aligned}\tag{3.3.25}$$

The last term can be handled similarly

$$\begin{aligned}
&\Big|\rho_2\int_0^1 q(x)\psi_t\int_0^{+\infty}g'(s)\eta_x^t(x,s)dsdx\Big|\\
&\le \rho_2\varepsilon_3\int_0^1\psi_t^2dx-\rho_2 g(0)C(\varepsilon_3)\int_0^1\int_0^{+\infty}g'(s)\mid\eta_x^t(x,s)\mid^2dsdx.
\end{aligned}\tag{3.3.26}$$

Combining (3.3.22)–(3.3.26), we obtain

$$\begin{aligned}
&\Big(\bar b\psi_x(0,t)+\int_0^{+\infty}g(s)\eta_x^t(0,s)ds\Big)^2+\Big(\bar b\psi_x(1,t)+\int_0^{+\infty}g(s)\eta_x^t(1,s)ds\Big)^2\\
&\le-\frac{d}{dt}\int_0^1\rho_2 q(x)\psi_t\Big(\bar b\psi_x+\int_0^{+\infty}g(s)\eta_x^t(x,s)ds\Big)dx\\
&\quad+\bar b^2\Big(4+\frac{1}{c_3}\Big)\int_0^1\psi_x^2dx+G_p\Big(4+\frac{1}{c_3}\Big)\int_0^1\int_0^{+\infty}g^p(s)\mid\eta_x^t(x,s)\mid^2dsdx\\
&\quad+2\rho_2\Big(\bar b+g_0+\frac{\varepsilon_3}{2}\Big)\int_0^1\psi_t^2dx+4k^2c_3\int_0^1(\varphi_x+\psi)^2dx\\
&\quad+4\delta^2c_3\int_0^1\theta_x^2dx-\rho_2 g(0)C(\varepsilon_3)\int_0^1\int_0^{+\infty}g'(s)\mid\eta_x^t(x,s)\mid^2dsdx.
\end{aligned}\tag{3.3.27}$$

Similarly, using (3.3.1), we arrive at

$$\frac{d}{dt}\int_0^1\rho_1 q\varphi_t\varphi_x dx\le -k\Big(\varphi_x^2(1)+\varphi_x^2(0)\Big)+3k\int_0^1\varphi_x^2dx+k\int_0^1\psi_x^2dx+2\rho_1\int_0^1\varphi_t^2dx.\tag{3.3.28}$$

Hence the assertion of the lemma follows from (3.3.18), (3.3.27) and (3.3.28).

Lemma 3.3.8 *Let $(\varphi, \psi, \theta, \eta^t)$ be the solution of problem (3.3.1)–(3.3.3), the functional I_4 defined by*

$$I_4 := -\rho_1 \int_0^1 \varphi_t \varphi dx - \rho_2 \int_0^1 \psi_t \psi dx$$

satisfies the estimate

$$\frac{dI_4}{dt} \leq -\rho_1 \int_0^1 \varphi_t^2 dx - \rho_2 \int_0^1 \psi_t^2 dx + (\bar{b} + 2\varepsilon_3) \int_0^1 \psi_x^2 dx + \frac{\delta^2}{4\varepsilon_3} \int_0^1 \theta_x^2 dx$$
$$+ k \int_0^1 (\varphi_x + \psi)^2 dx + \frac{G_p}{4\varepsilon_3} \int_0^1 \int_0^{+\infty} g^p(s) \mid \eta_x^t(x,s) \mid^2 dsdx, \quad (3.3.29)$$

for any $\varepsilon_3 > 0$.

Proof It is easy to check that

$$\frac{dI_4}{dt} = -\rho_1 \int_0^1 \varphi_t^2 dx - \rho_2 \int_0^1 \psi_t^2 dx + \bar{b} \int_0^1 \psi_x^2 dx + \delta \int_0^1 \theta_x \psi dx$$
$$+ k \int_0^1 (\varphi_x + \psi)^2 dx + \int_0^1 \psi_x \int_0^{+\infty} g(s) \eta_x^t(x,s) dsdx, \quad (3.3.30)$$

and from Lemma 3.3.3 and Young's inequality, (3.3.29) follows.

Now for $N, N_1, N_2, \lambda > 0$, we define a Lyapunov functional $\mathcal{F}(t)$ as follows

$$\mathcal{F}(t) = NE(t) + N_1 I_1(t) + N_2 I_2(t) + I_3(t) + \lambda I_4(t) + \frac{\varepsilon_3}{k} \int_0^1 \rho_1 q \varphi_t \varphi_x dx$$
$$+ \frac{1}{4\varepsilon_3} \int_0^1 \rho_2 q(x) \psi_t \Big(\bar{b} \psi_x + \int_0^{+\infty} g(s) \eta_x^t(x,s) ds \Big) dx. \quad (3.3.31)$$

Using (3.3.7), (3.3.11), (3.3.14), (3.3.16), (3.3.17) and (3.3.29), we get

$$\frac{d}{dt}\mathcal{F}(t) \leq \Upsilon_1 \int_0^1 \psi_x^2 dx + \Upsilon_2 \int_0^1 \varphi_t^2 dx + \Upsilon_3 \int_0^1 \psi_t^2 dx + \Upsilon_4 \int_0^1 (\varphi_x + \psi)^2 dx$$
$$+ \Upsilon_5 \int_0^1 \theta_x^2 dx + C_2 \int_0^1 \int_0^{+\infty} g^p(s) \mid \eta_x^t(x,s) \mid^2 dsdx$$
$$+ (N - C_1) \int_0^1 \int_0^{+\infty} g'(s) \mid \eta_x^t(x,s) \mid^2 dsdx, \quad (3.3.32)$$

where C_1, C_2 are positive constants independent of N and

$$\begin{cases} \Upsilon_1 = -N_1(\bar{b} - 2c_1) + N_2\varepsilon_2\bar{b} + C(c_3, \lambda, \varepsilon_3), \\ \Upsilon_2 = N_1\varepsilon_1\rho_1 + \varepsilon_3 + \dfrac{2\rho_1\varepsilon_3}{k} - \lambda\rho_1, \\ \Upsilon_3 = N_1(\rho_2 + \dfrac{\rho_1}{4\varepsilon_1}) - N_2\dfrac{g_0\rho_2}{2} + \rho_2 + \dfrac{1}{4\varepsilon_3} - \lambda\rho_2, \\ \Upsilon_4 = -k + \dfrac{k^2c_3}{\varepsilon_3} + N_2\varepsilon_2k^2 + 2(3\varepsilon_3 + \lambda k), \\ \Upsilon_5 = -N\beta + \dfrac{1}{4c_1} + \varepsilon_2\delta^2c_3. \end{cases}$$

Now we choose our constants very carefully and properly so that there exists a constant $\mu > 0$, such that the following form holds

$$\frac{d}{dt}\mathcal{F}(t) \le -\mu\bigg(\int_0^1 \Big(\psi_x^2 + \psi_t^2 + \varphi_t^2 + \theta_x^2 + (\varphi_x + \psi)^2\Big)dx + \int_0^1\int_0^{+\infty} g^p(s) \mid \eta_x^t(x,s) \mid^2 dsdx\bigg). \tag{3.3.33}$$

On the other hand, we know that $\mathcal{F}(t)$ is equivalent to $E(t)$, i.e.,

$$\mathcal{F}(t) \sim E(t). \tag{3.3.34}$$

We are now ready to prove Theorem 3.3.1.

Proof of Theorem 3.3.1 We distinguish two cases.
Case 1 $p = 1$. From (3.3.33) and (3.3.34), it follows

$$\frac{d}{dt}\mathcal{F}(t) \le -\omega\mathcal{F}(t). \tag{3.3.35}$$

Integrating (3.3.35) over $(0, t)$, and using (3.3.34), (3.3.5) follows immediately.
Case 2 $p > 1$. Firstly, from (3.3.4) and (3.3.7), we derive

$$\begin{aligned} E^{2p-1}(t) &\le C(E(0))^{2p-2}\int_0^1 \left(\psi_x^2 + \psi_t^2 + \varphi_t^2 + \theta_x^2 + (\varphi_x + \psi)^2\right)dx \\ &\quad + C\int_0^1\left(\int_0^{+\infty} g(s) \mid \eta_x^t(x,s) \mid^2 ds\right)^{2p-1} dx \\ &\le C(E(0))^{2p-2}\int_0^1 \left(\psi_x^2 + \psi_t^2 + \varphi_t^2 + \theta_x^2 + (\varphi_x + \psi)^2\right)dx \\ &\quad + C\int_0^1\int_0^{+\infty} g^p(s) \mid \eta_x^t(x,s) \mid^2 dsdx. \end{aligned} \tag{3.3.36}$$

Combining (3.3.33), (3.3.34) and (3.3.36), we have

$$\mathcal{F}'(t) \le -CE^{2p-1}(t) \le -C\mathcal{F}^{2p-1}(t),$$

which, integrated over $(0, t)$, leads to

$$\mathcal{F}(t) \le \frac{C}{(t+1)^{1/(2p-2)}}. \tag{3.3.37}$$

We observe that

$$\begin{aligned}
&\int_0^1 \int_0^{+\infty} g(s) \mid \eta_x^t(x,s) \mid^2 dsdx \\
&\le \left(\int_0^1 \int_0^{+\infty} \mid \eta_x^t(s) \mid^2 dsdx \right)^{(p-1)/p} \times \left(\int_0^1 \int_0^{+\infty} g^p(s) \mid \eta_x^t(x,s) \mid^2 dsdx \right)^{1/p}
\end{aligned} \tag{3.3.38}$$

and use (3.3.4) and (3.3.37) to conclude, for $p < 3/2$,

$$\begin{aligned}
\int_0^1 \int_0^t \mid \eta_x^t(x,s) \mid^2 dsdx &= 2 \int_0^1 \int_0^{+\infty} \mid \psi_{x,t}(x,s) \mid^2 dsdx \\
&\quad + 2 \int_0^1 \int_0^{+\infty} \mid \psi_x(x, t-s) \mid^2 dsdx \\
&\le 2t \int_0^1 \mid \psi_x(x,t) \mid^2 dx + \frac{4}{\bar{b}} \int_0^t E(t-s)ds \\
&\le \frac{Ct}{(t+1)^{1/(2p-2)}} + C \int_0^t \frac{ds}{(t-s+1)^{1/(2p-2)}} \\
&\le \Lambda,
\end{aligned}$$

where $\Lambda > 0$ is a constant independent of t. Hence, we get

$$\int_0^1 \int_0^{+\infty} \mid \eta_x^t(x,s) \mid^2 dsdx \le \Lambda. \tag{3.3.39}$$

Consequently, a combination of (3.3.38) and (3.3.39) gives

$$\int_0^1 \int_0^t \mid g(s)\eta_x^t(x,s) \mid^2 dsdx \le \Lambda^{(p-1)/p} \left(\int_0^1 \int_0^{+\infty} g^p(s) \mid \eta_x^t(x,s) \mid^2 dsdx \right)^{1/p}$$

or

$$\left(\int_0^1 \int_0^t \mid g(s)\eta_x^t(x,s) \mid^2 dsdx \right)^p \le C \int_0^1 \int_0^{+\infty} g^p(s) \mid \eta_x^t(x,s) \mid^2 dsdx.$$

Similarly to (3.3.36), we obtain

$$E^p(t) \le C \int_0^1 \left(\psi_x^2 + \psi_t^2 + \varphi_t^2 + \theta_x^2 + (\varphi_x + \psi)^2\right) dx$$
$$+ C \int_0^1 \int_0^{+\infty} g^p(s) \mid \eta_x^t(s) \mid^2 dsdx \tag{3.3.40}$$

which, together with (3.3.33), (3.3.34) and (3.3.40), yields

$$\mathcal{F}'(t) \le -C\mathcal{F}^p(t). \tag{3.3.41}$$

Integrating (3.3.41) gives

$$\mathcal{F}(t) \le \frac{C}{(t+1)^{1/(p-1)}}, \tag{3.3.42}$$

which, along with (3.3.34), yields (3.3.6). This completes the proof of Theorem 3.3.1. □

3.4 Case of Nonequal Wave-Speeds

In this section, we treat the case of different wave speeds propagation ($\frac{\rho_1}{k} \neq \frac{\rho_2}{b}$), we shall show that if the initial data are regular enough, the solution energy $E(t)$ decays in the rate of $1/t$ when the relaxation function g decays exponentially.

Define the second-order energy as

$$E_1(t) = E(\varphi_t, \psi_t, \theta_t, \eta_t^t),$$

where E is given in (3.3.4).

Theorem 3.4.1 *Let* $(\varphi_0, \varphi_1), (\psi_0, \psi_1) \in (H^2(0,1) \cap H_0^1(0,1)) \times H_0^1(0,1), \theta_0 \in H_0^1(0,1)$ *and* $\eta_0^t \in L_g^2(\mathbb{R}^+, H^2(0,1) \cap H_0^1(0,1))$. *Then there exists a constant* $C > 0$ *such that for all* $t > 0$,

$$E(t) \le C/t, \quad p = 1. \tag{3.4.1}$$

In order to prove the above theorem, we need the following two lemmas.

Lemma 3.4.1 *Let* $(\varphi, \psi, \theta, \eta^t)$ *be the solution of problem (3.3.1)–(3.3.3), then we have*

$$E'(t) = -\beta \int_0^1 \theta_{tx}^2 dx + \frac{1}{2}\int_0^1 \int_0^{+\infty} g'(s) \mid \eta_{tx}^t(x,s) \mid^2 dsdx \le 0. \tag{3.4.2}$$

Proof Similarly to Lemma 3.3.1, we can prove the lemma.

Lemma 3.4.2 *Let $(\varphi, \psi, \theta, \eta^t)$ be the solution of problem (3.3.1)–(3.3.3), then for any $\varepsilon_3 > 0$, we conclude*

$$\begin{aligned}\frac{dI_3(t)}{dt} \leq & \left(\varphi_x\left(\bar{b}\psi_x + \int_0^{+\infty} g(s)\eta_x^t(x,s)\right)\right)\Big|_{x=0}^{x=1} - k\int_0^1 (\varphi_x+\psi)^2 dx \\ & + \rho_2\int_0^1 \psi_t^2 dx + \varepsilon_3\int_0^1 \varphi_t^2 dx + \frac{1}{4\varepsilon_3}\int_0^1 \theta_x^2 dx \\ & - g(0)C(\varepsilon_3)\int_0^1\int_0^{+\infty} g'(s) \mid \eta_x^t(x,s)\mid^2 dsdx \\ & + C(\varepsilon_3)\int_0^1\int_0^{+\infty} g^p(s)\mid \eta_{tx}^t(x,s)\mid^2 dsdx. \qquad (3.4.3)\end{aligned}$$

Proof Similarly to the proof of Lemma 3.3.6, we obtain

$$\begin{aligned}\frac{dI_3(t)}{dt} = & \int_0^1 (\varphi_x+\psi)\left(\bar{b}\psi_{xx} + \int_0^{+\infty} g(s)\eta_{xx}^t(x,t-s)ds - k(\varphi_x+\psi) - \delta\theta_x\right)dx \\ & + \rho_2\int_0^1 \psi_t^2 dx + \bar{b}\int_0^1 \psi_x(\varphi_x+\psi)_x dx + \left(\frac{\rho_1 b}{k} - \rho_2\right)\int_0^1 \psi_{tx}\varphi_t dx \\ & - \frac{\rho_1}{k}\int_0^1 \varphi_t\int_0^{+\infty} g(s)\eta_{sx}^t(x,s)dsdx + \int_0^1 (\varphi_x+\psi)_x\int_0^{+\infty} g(s)\eta_x^t(x,s)dsdx.\end{aligned}$$

Since $\frac{\rho_1}{k} \neq \frac{\rho_2}{b}$, we have to handle the term $\int_0^1 \psi_{tx}\varphi_t dx$ on the right-hand side of the above equation. To this end, we use the idea of [51] to obtain

$$\int_0^1 \psi_{tx}\varphi_t dx = \frac{1}{g_0}\int_0^1\left(\int_0^{+\infty} g(s)\eta_{tx}^t(x,s)ds\right)\varphi_t dx - \frac{1}{g_0}\int_0^1\left(\int_0^{+\infty} g'(s)\eta_x^t(x,s)ds\right)\varphi_t dx. \qquad (3.4.4)$$

Therefore, using Young's inequality and (3.3.9), we obtain the following two estimates

$$\begin{aligned}& \frac{1}{g_0}\left(\frac{\rho_1 b}{k} - \rho_2\right)\int_0^1\left(\int_0^{+\infty} g(s)\eta_{tx}^t(x,s)ds\right)\varphi_t dx \\ & \leq \frac{\varepsilon_3}{2}\int_0^1 \varphi_t^2 dx + C(\varepsilon_3)\int_0^1\int_0^{+\infty} g^p(s)\mid \eta_{tx}^t(x,s)\mid^2 dsdx \qquad (3.4.5)\end{aligned}$$

and

$$\begin{aligned}& \frac{1}{g_0}\left(\frac{\rho_1 b}{k} - \rho_2\right)\int_0^1\left(\int_0^{+\infty} g'(s)\eta_{tx}^t(x,s)ds\right)\varphi_t dx \\ & \leq \frac{\varepsilon_3}{2}\int_0^1 \varphi_t^2 dx - g(0)C(\varepsilon_3)\int_0^1\int_0^{+\infty} g'(s)\mid \eta_{tx}^t(x,s)\mid^2 dsdx. \qquad (3.4.6)\end{aligned}$$

Inserting (3.4.5) and (3.4.6) into (3.4.4), using the estimates of Lemma 3.3.7, we obtain the desired result (3.4.3). □

Proof of Theorem 3.4.1 We define a Lyapunov functional $\mathcal{F}$ as follows

$$\mathcal{F}(t) = N\Big(E(t) + E_1(t)\Big) + N_1 I_1 + N_2 I_2 + I_3 + \lambda I_4 + \frac{\varepsilon_3}{k}\int_0^1 \rho_1 q\varphi_t\varphi_x dx$$
$$+\frac{1}{4\varepsilon_3}\int_0^1 \rho_2 q(x)\psi_t\left(\bar{b}\psi_x + \int_0^{+\infty} g(s)\eta_x^t(x,s)ds\right)dx. \tag{3.4.7}$$

Consequently, we get

$$\frac{d}{dt}\mathcal{F}(t) \le \Upsilon_1\int_0^1 \psi_x^2 dx + \hat{\Upsilon}_2\int_0^1 \varphi_t^2 dx + \Upsilon_3\int_0^1 \psi_t^2 dx + \Upsilon_4\int_0^1 (\varphi_x+\psi)^2 dx$$
$$+ \Upsilon_5\int_0^1 \theta_x^2 dx + C_2\int_0^1\int_0^{+\infty} g^p(s)\mid \eta_x^t(x,s)\mid^2 dsdx$$
$$+ (\frac{N}{2} - C_1)\int_0^1\int_0^{+\infty} g'(s)\mid \eta_x^t(x,s)\mid^2 dsdx$$
$$+ \frac{N}{2}\int_0^1\int_0^{+\infty} g'(s)\mid \eta_{tx}^t(x,s)\mid^2 dsdx$$
$$+ C(\varepsilon_3)\int_0^1\int_0^{+\infty} g^p(s)\mid \eta_{tx}^t(x,s)\mid^2 dsdx,$$

where

$$\hat{\Upsilon}_2 = \Upsilon_2 + \varepsilon_3.$$

Similarly to the proof of (3.3.33), there is a constant $\mu_1 > 0$ such that

$$\frac{d}{dt}\mathcal{F}(t) \le -\mu_1\Bigg(\int_0^1 (\psi_x^2 + \psi_t^2 + \varphi_t^2 + \theta_x^2 + (\varphi_x+\psi)^2)dx$$
$$+\int_0^1\int_0^{+\infty} g^p(s)\mid \eta_x^t(x,s)\mid^2 dsdx + \int_0^1\int_0^{+\infty} g^p(s)\mid \eta_{tx}^t(x,s)\mid^2 dsdx\Bigg). \tag{3.4.8}$$

Then, for $p = 1$, it is not hard to see that

$$\frac{d}{dt}\mathcal{F}(t) \le -cE(t). \tag{3.4.9}$$

Integrating (3.4.9) and noticing (3.4.7), we get

$$\int_0^t E(s)ds \le C\Big(\mathcal{F}(0) - \mathcal{F}(t)\Big) \le C\mathcal{F}(0) \le C\Big(E(0) + E_1(0)\Big), \quad for \ \ all \ t > 0. \tag{3.4.10}$$

Noting that

$$\frac{d}{dt}\Big(tE(t)\Big) = E(t) + t\frac{d}{dt}E(t) \le E(t),$$

we have

$$tE(t) \le \int_0^t E(s)ds \le c\Big(E(0) + E_1(0)\Big), \quad for \ \ all \ t > 0.$$

Consequently, we get

$$E(t) \le \frac{C}{t}, \quad for \ \ all \ t > 0. \tag{3.4.11}$$

□

3.5 Bibliographic Comments

Timoshenko [270] first proposed the transverse vibrations of a beam which are given by the following coupled partial differential equations

$$\begin{cases} \rho u_{tt} = (k(u_x - \varphi))_x, & \text{in} \quad (0, L) \times (0, +\infty), \\ I_\rho \varphi_{tt} = (EI\varphi_x)_x + k(u_x - \varphi), & \text{in} \quad (0, L) \times (0, +\infty), \end{cases} \tag{3.5.1}$$

where t denotes the time variable and x is the space variable along the beam, the length of which is L, in its equilibrium configuration. The function u is the transverse displacement of the beam and φ is the rotation angle of the filament of the beam. The coefficients ρ, I_ρ, E, I and k are respectively the mass per unit length, the polar moment of the inertia of a cross section, Young's modulus of elasticity, the moment of inertia of a cross section, and the shear modulus. If we denote $\rho_1 = \rho$, $\rho_2 = I_\rho, b = EI$, and we derive directly from (3.5.1) the following system

$$\begin{cases} \rho_1 u_{tt} = (k(u_x - \varphi))_x, & \text{in} \quad (0, L) \times (0, +\infty), \\ \rho_2 \varphi_{tt} = (b\varphi_x)_x + k(u_x - \varphi), & \text{in} \quad (0, L) \times (0, +\infty). \end{cases}$$

The energy of the beam is given by

$$E(t) = \frac{1}{2}\int_0^L \left(\rho_1|u_t|^2 + \rho_2|\varphi_t|^2 + b|\varphi_t|^2 + k|u_x - \varphi|^2\right) dx.$$

There are many results on global existence and asymptotic behavior of solutions to the system (3.5.1) (see, e.g., [128, 192, 254, 260, 267, 280]). Let us here mention some results on the stability of this model and the Timoshenko systems in thermoelasticity. The case of two boundary control forces has already been considered by Kim and Renardy [128] for the Timoshenko beam, who proved the exponential decay of the energy $E(t)$ by using a multiplier technique and provided numerical estimates of the eigenvalues of the operator associated to this system, and by Lagnese and Lions [136], for the study of the exact controllability. Later on, Yan [280] generalized the result of [128] by considering nonlinear boundary conditions of the form

$$\begin{cases} k(\varphi(L,t) - u_x(L,t)) = f_1(u_t(L,t)), \quad for \;\; all\, t > 0, & (3.5.2) \\ -EI\varphi_x(L,t) = f_2(\varphi_t(L,t)), \quad for \;\; allt > 0, & (3.5.3) \end{cases}$$

where f_1, f_2 are functions with a polynomial growth near the origin.

Muñoz Rivera and Racke [190] considered a nonlinear Timoshenko system as an initial-boundary value problem, the system has a dissipative mechanism through frictional damping being present only in the equation for the rotation angle, they gave an alternative proofs for a sufficient and necessary condition for exponential stability for the linear case, and proved the exponential stability for the nonlinear case. Shi and Feng [254] established the exponential decay of the energy with two locally distributed feedbacks. Later on, this result was improved by Soufyane and Wehbe [260], who showed that it is possible to stabilize (3.5.1) uniformly by using a unique locally distributed feedback, in fact, they proved that the uniform stability holds if and only if the wave speeds are equal; otherwise only the asymptotic stability has been proved. Grasselli and Pata [204] considered the Timoshenko model of a viscoelastic beam consisting of two coupled second order linear integrodifferential hyperbolic equations with semilinear external forces acting on the beam. Then the authors proved that the corresponding solution decays exponentially to zero, as time goes to infinity, provided the relaxation kernels have certain dissipative properties and also decay exponentially to zero. The decay rate is explicitly calculated, which allows the authors to show the existence of a uniform attractor, provided the semi-linear external forces satisfy appropriate conditions. The method used is based on semigroup theory. To apply it, they introduced new variables representing the past histories of the solution. Grobbelaar-Van Dalsen [93] considered a hybrid elastic model consisting of a Timoshenko beam and a tip load at the free end of the beam. He showed that uniform stabilization of the model which includes the rotary inertia of the tip load can be obtained when feedback boundary moment and force controls are applied at the point of contact between the beam and the tip load. However, in the presence of the load stabilization is slower and subject to a restriction on the boundary data at the free end of the beam.

Ammar-Khodja et al. [4] proved the exponential decay of the associated energy for Timoshenko systems with memory, further used the multiplier techniques to prove that the system is uniformly stable if and only if the wave speeds are equal ($\frac{k}{\rho_1} = \frac{b}{\rho_2}$) and g decays uniformly. Precisely, they proved an exponential decay if g decays in an exponential rate and polynomial decay if g decays in a polynomial rate. They also required some extra technical conditions on both g' and g'' to obtain their results. Recently, this result was improved by Messaoudi and Mustafa [172], for a wider class of relaxation functions and without imposing any restrictive growth assumption on the damping term, who established an explicit and more general decay rate result, from which the exponential and the polynomial decay result are only special cases. Later on, Fernández Sare and Muñoz Rivera [80] considered a similar Timoshenko-type system with a past history of the form

$$\begin{cases} \rho_1\varphi_{tt} - k(\varphi_x + \psi)_x = 0, & (3.5.4)\\ \rho_2\psi_{tt} - b\psi_{xx} + \int_0^{+\infty} g(s)\psi_{xx}(t-s)ds + k(\varphi_x + \psi) = 0, & (3.5.5) \end{cases}$$

where ρ_1, ρ_2, k, b are positive constants and g is a positive twice differentiable function satisfying, for some constants $k_0, k_1, k_2 > 0$,

$$\begin{cases} g(t) > 0, \quad -k_0 g(t) \leq g'(t) \leq -k_1 g(t), \quad | g''(t) | \leq k_2(t), \quad for \ all \ t > 0, & (3.5.6)\\ \hat{b} = b - \int_0^{+\infty} g(s)ds > 0, & (3.5.7) \end{cases}$$

and showed that the dissipation given by the history term is strong enough to stabilize the system exponentially if and only if the wave speeds are equal, and further proved that the solution decays polynomially for the case of different wave speeds. Racke and Said-Houari [237] studied the asymptotic behavior of a nonlinear version of the Timoshenko system in thermoelasticity with second sound. As it has been already proved in Said-Houari and Kasimov [247], the linear version of this system is of regularity-loss type. It is well known that the regularity-loss property of the linear problem creates difficulties when dealing with the nonlinear problem. In fact, the dissipative property of the problem becomes very weak in the high frequency region and as a result the classical energy method fails. To overcome this difficulty, they used an energy method with negative weights to create an artificial damping which allows them to control the nonlinearity. Grasselli, Muñoz Rivera and Pata analyzed the longterm properties of a C_0-semigroup describing the solutions to a linear evolution system that models a thermoelastic thin plate with memory effects in the heat flux law. Under reasonably general assumptions on the memory kernel, all single trajectories are shown to decay to zero. In spite of that, the semigroup is not exponentially stable.

Taylor [267] studied the boundary control of the system (3.5.1) with variable physical characteristics. Raposo et al. [242] studied the system with homogeneous Dirichlet boundary conditions and two linear frictional damping and proved that the energy decays exponentially. This result is similar to that in [267]. As the authors mentioned, their analysis is based on the method developed by Zheng and Liu and their collaborators in past years, their work lies on the semigroup theory method, which was developed by Liu and Zheng [155]. Cavalcanti et al. [31] considered the Timoshenko model for vibrating beams under effect of two nonlinear and localized frictional damping mechanisms acting on the transverse displacement and on the rotational angle. They proved that the damping placed on an arbitrarily small support, unquantitized at the origin and without assuming equal speeds of propagation of waves, leads to uniform decay rates (asymptotic in time) for the energy function. This result removes the necessity (as long as both transverse displacements and rotational angles are minimally damped) of the assumption on equal speeds. Recently, Bassam et al. [11] studied the indirect boundary stabilization of the Timoshenko system with only one dissipation law. This system, which models the dynamics of a beam, is a hyperbolic system with two wave speeds. Assuming that the wave speeds are equal, they proved non-uniform stability and an optimal polynomial energy decay rate is obtained. Otherwise, if the ratio of the wave speeds is a rational number, they showed that the decay rate is of polynomial type. Xu et al. [279] concerned the Timoshenko system in the case of non-equal wave speeds, which admits the dissipative structure of regularity-loss. Firstly, they constructed global solutions to the Timoshenko system pertaining to data in the Besov space with the regularity $s = 3/2$. Owing to the weaker dissipative mechanism, extra higher regularity than that for the global-in-time existence is usually imposed to obtain the optimal decay rates of classical solutions, so it is almost impossible to obtain the optimal decay rates in the critical space. To overcome the outstanding difficulty, they developed a new frequency-localization time-decay inequality, which captures the information related to the integrability at the high-frequency part. Furthermore, by the energy approach in terms of high-frequency and low-frequency decomposition, they showed the optimal decay rate for Timoshenko system in critical Besov spaces. Muñoz Rivera and Avila [187] considered the uniform stabilization of a hybrid elastic model consisting of a Timoshenko beam and a tip load at the free end of the beam, the authors used semigroup approach and proved that the semigroup e^{At} associated to this model is not exponentially stable. Moreover, they proved that the semigroup decays polynomially to zero as $t^{-1/2}$. When the damping mechanism is effective only on the boundary of the rotational angle, the solution also decays polynomially as $t^{-1/2}$ provided the wave speeds are equal. Otherwise it decays as $t^{-1/4}$for any initial data taken in $D(A)$. Dell'Oro and Pata [55] analyzed the decay properties of the solution semigroup generated by the linear Timoshenko system in presence of a flat (i.e., piecewise constant) memory kernel μ. In this situation, the uniform decay of the solutions does not occur, also when the two equations exhibit the same propagation speed. Guesmia and Messaoudi [99] considered a vibrating system of Timoshenko-type in a one-dimensional bounded domain with complementary frictional damping and infinite memory acting on the transversal displacement. They showed that the dissipation generated by these two complementary controls guarantees the stability of the system in case of the equal-speed propagation as well as in the opposite case. They established in each case a general decay estimate of the solutions. In the particular case when the wave propagation speeds are different and the frictional damping is linear, they gave a relationship between the smoothness of the initial data and the decay rate of the solutions.

For the Mindlin–Timoshenko models, Araruna, Silva and Zuazua [7] showed how the so-called von Karman model can be obtained as a singular limit of a modified Mindlin–Timoshenko system when the modulus of elasticity in shear k tends to infinity, provided a regularizing term through a fourth order dispersive operator is added. Introducing damping mechanisms, the authors also showed that the energy of solutions for this modified Mindlin–Timoshenko system decays exponentially, uniformly with respect to the parameter k. As $k \to +\infty$, the authors obtained the damped von Karman model with associated energy exponentially decaying to zero as well. Fastovska [70] considered a nonlinear Mindlin–Timoshenko system for a thermoelastic plate with heat conduction of Gurtin-Pipkin type. The main result was that under certain assumptions on the nonlinearities the long-time behavior of the system in the limit case as the relaxation time tends to zero can be described by the classical Mindlin–Timoshenko thermoelasticity model. In this sense, the author proved the existence of a compact global attractor whose fractal dimension is finite and the upper semicontinuity of the attractor as the relaxation time tends to zero. Later on, the the same author [71] considered a nonlinear problem for a thermoviscoelastic Mindlin–Timoshenko plate with hereditary heat conduction is considered here. They proved the existence of a compact global attractor whose fractal dimension is finite. The main aim of this work is to show the upper semicontinuity of the attractor as the relaxation kernels fade in a suitable sense. Recently, Grobbelaar-Van Dalsen [96] was concerned with the polynomial stabilization of a two-dimensional thermoelastic Mindlin–Timoshenko plate model with no mechanical damping. By applying a frequency domain method in which an inequality was used for the trace of Sobolev functions on the boundary of a bounded, open connected set $\Omega \subset \mathbb{R}^n$, $(n \geq 2)$, they showed that the decay is slower than in the model considered in [95]. Chueshov and Lasiecka [38] considered dynamics of a class of Mindlin–Timoshenko plate models with nonlinear feedback forces and proved the existence of a compact global attractor and studied its limiting properties when the shear modulus tends to infinity.

For the Timoshenko systems in classical thermoelasticity, Messaoudi et al. [170, 174, 175], Muñoz Rivera and Racke [191] had done a lot of works. Several exponential decay results for both linear and nonlinear cases have been established. Also Messaoudi and Said-Houari [175] considered a Timoshenko-type system of thermoelasticity of type III and proved an exponential decay similar to the one in [174, 191]. In particular, they also proved the energy decay in a Timoshenko-type system with history in thermoelasticity of type III in [170] by the energy method. At the same time, in the next chapter, Qin and Ma, the authors of the present book considered the same model as in [170] and established a polynomial stability result for the case of nonequal wave speed, but using different methods namely semigroup method. Fatori et al. [76] studied a fully hyperbolic thermoelastic Timoshenko system with past history where the thermal effects are given by Cattaneo's law, and obtained the exponential stability of solutions if and only if a new condition on the wave speed of propagation is verified, otherwise, when that condition fails, they obtained polynomial stability of solutions. In this chapter, using the method in [170], we establish a similar result of the energy decay in a Timoshenko-type system with memory in thermoelasticity of type I. In addition, Qin and Ma, the authors of the present book also considered the same model as in [159] by using the semigroup method to prove the polynomial stability result with assumptions on past history relaxation function g exponentially decaying for the nonequal wave-speed case.

Last, we would like to mention other works in [3, 6, 78, 79, 151, 171, 175, 202, 270] for related models.

Chapter 4
Stability for a Timoshenko-type Thermoelastic Equations of Type III with a Past History

4.1 Introduction

In this chapter, we shall use the semigroup method to establish a polynomial stability result for the case of nonequal wave speed to the following system

$$\begin{cases} \rho_1\varphi_{tt} - k(\varphi_x + \psi)_x = 0, \\ \rho_2\psi_{tt} - b\psi_{xx} + \int_0^{+\infty} g(s)\psi_{xx}(t-s)ds + k(\varphi_x + \psi) + \delta\theta_{tx} = 0, \\ \rho_3\theta_{tt} - \beta\theta_{txx} - \bar{\beta}\theta_{xx} + \delta\psi_{xt} = 0 \end{cases} \tag{4.1.1}$$

with positive constants $\rho_1, \rho_2, \rho_3, k, b, \beta, \delta$ together with initial conditions

$$\varphi(\cdot,0) = \varphi_0,\ \varphi_t(\cdot,0) = \varphi_1,\ \psi(\cdot,0) = \psi_0,\ \psi_t(\cdot,0) = \psi_1,\ \theta(\cdot,0) = \theta_0 \tag{4.1.2}$$

and boundary conditions

$$\varphi_x(0,t) = \varphi_x(1,t) = \psi(0,t) = \psi(1,t) = \theta(0,t) = \theta(1,t) = 0, \tag{4.1.3}$$

where the functions φ, ψ and θ depend on $(x,t) \in [0,1] \times [0,+\infty)$ and model the transverse displacement of a beam with reference configuration $(0,1) \subset \mathbb{R}$, the rotation angle of a filament and the temperature difference respectively.

Throughout this chapter, we use the notation "$\to 0$" to denote the strong convergence to zero in the corresponding space. We adopt the results in this chapter from [162]. This model is similar to that in Chap. 3, but they are different, one is Timoshenko-type thermoelastic equations of type I, another one is Timoshenko-type thermoelastic equations of type III. At the same time, we use the different methods to obtain our results.

Y. Qin and Z. Ma, *Global Well-posedness and Asymptotic Behavior of the Solutions to Non-classical Thermo(visco)elastic Models*,
DOI 10.1007/978-981-10-1714-8_4

4.2 Preliminaries

In order to state our main result, we make the following hypotheses:
$g : \mathbb{R}^+ \to \mathbb{R}^+$ is a differentiable function such that there exist constants $k_0, k_1 > 0$ satisfying

$$g(0) > 0,\ k_0 g(t) \leq g'(t) \leq -k_1 g(t), \mid g''(t) \mid \leq k_2 g(t), \tag{4.2.1}$$

$$b - \int_0^{+\infty} g(s)ds = \bar{b} > 0. \tag{4.2.2}$$

We introduce

$$\eta^t(x, s) = \psi(x, t) - \psi(x, t - s), \quad t, s \geq 0, \tag{4.2.3}$$

then we can reformulate system (4.1.1) as the system

$$\begin{cases} \rho_1 \varphi_{tt} - k(\varphi_x + \psi)_x = 0, \\ \rho_2 \psi_{tt} - \bar{b}\psi_{xx} - \int_0^{+\infty} g(s)\eta^t_{xx}(t - s)ds + k(\varphi_x + \psi) + \delta\theta_{tx} = 0, \\ \rho_3 \theta_{tt} - \bar{\beta}\theta_{txx} - \bar{\beta}\theta_{xx} + \delta\psi_{xt} = 0, \\ \eta^t_t + \eta^t_s - \psi_t = 0 \end{cases} \tag{4.2.4}$$

with initial conditions

$$\varphi(\cdot, 0) = \varphi_0,\ \varphi_t(\cdot, 0) = \varphi_1,\ \psi(\cdot, 0) = \psi_0,\ \psi_t(\cdot, 0) = \psi_1,\ \theta(\cdot, 0) = \theta_0, \eta^0(x, s) = \eta_0 \tag{4.2.5}$$

and boundary conditions

$$\varphi_x(0, t) = \varphi_x(1, t) = \psi(0, t) = \psi(1, t) = \theta(0, t) = \theta(1, t) = 0. \tag{4.2.6}$$

We shall use the semigroup method to establish the exponential stability. To this end, we can rewrite the system (4.2.4) as an evolution equation

$$\begin{cases} U = (\varphi, \varphi_t, \psi, \psi_t, \theta, \theta_t, \eta)^T \equiv (u_1, u_2, u_3, u_4, u_5, u_6, u_7), \\ U_t = \mathcal{A}U, \quad U(0) = U_0 \end{cases}$$

where $U_0 = (\varphi_0, \varphi_1, \psi_0, \psi_1, \theta_0, \theta_1, \eta_0)$, and $\mathcal{A}$ is the differential operator

$$\mathcal{A} = \begin{pmatrix} 0 & Id & 0 & 0 & 0 & 0 & 0 \\ \frac{k}{\rho_1}\partial_x^2 & 0 & \frac{k}{\rho_1}\partial_x & 0 & 0 & 0 & 0 \\ 0 & 0 & 0 & Id & 0 & 0 & 0 \\ -\frac{k}{\rho_2}\partial_x & 0 & \frac{\bar{b}}{\rho_2}\partial_x^2 - \frac{k}{\rho_2}Id & 0 & 0 & -\frac{\delta}{\rho_2}\partial_x & -\frac{1}{\rho_2}\int_0^{+\infty} g(s)\partial_x^2(s)ds \\ 0 & 0 & 0 & -\frac{\delta}{\rho_3} & \frac{\bar{\beta}}{\rho_3}\partial_x^2 & \frac{\beta}{\rho_3}\partial_x^2 & 0 \\ 0 & 0 & 0 & Id & 0 & 0 & -\partial_s \end{pmatrix}.$$

Let

$$\mathcal{H} = H_*^1(0,1) \times L_*^2(0,1) \times H_0^1(0,1) \times L^2(0,1) \times H_0^1(0,1) \times L^2(0,1) \times L_g^2(\mathbb{R}, H_0^1)$$

with

$$L_*^2(0,1) := \left\{ v \in L^2(0,1) : \int_0^1 v(x)dx = 0 \right\},$$

$$H_*^1(0,1) := \left\{ v \in H^1(0,1) : \int_0^1 v(x)dx = 0 \right\}$$

and norm given by

$$\begin{aligned} \| U \|_{\mathcal{H}}^2 = \rho_1 \| u^2 \|_{L^2}^2 + \rho_2 \| u^4 \|_{L^2}^2 + k \| u_x^1 + u^3 \|_{L^2}^2 + \rho_3 \| u^6 \|_{L^2}^2 \\ + \bar{b} \| u_x^3 \|_{L^2}^2 + \rho_3 \| u_x^5 \|_{L^2}^2 + \rho_1 \| u^7 \|_{L_g^2(\mathbb{R}^+, H_0^1)}^2 \end{aligned}$$

is a Hilbert space, where $L_g^2(\mathbb{R}^+, H_0^1)$ denotes the Hilbert space of H_0^1-valued functions on $\mathbb{R}^+$, endowed with the inner product

$$< \varphi, \psi >_{L_g^2(\mathbb{R}^+, H_0^1)} = \int_0^1 \int_0^{+\infty} g(s)\varphi_x(x,s)\psi_x(x,s)dsdx.$$

The domain of the operator $\mathcal{A}$ is defined by

$$\begin{aligned} D(\mathcal{A}) = \Big\{ & U = (u^1, u^2, u^3, u^4, u^5, u^6, u^7) \in \mathcal{H} \mid u^1 \in H^2(0,1), u_x^1 \in H_0^1(0,1), \\ & u^2 \in H_*^1(0,1), u^4 \in H_0^1(0,1), u_x^5 \in H^2(0,1) \cap H_0^1(0,1), u_s^7 \in L_g^2(\mathbb{R}^+, H_0^1), \\ & \bar{b}u^3 + \int_0^{+\infty} g(s)u^7(x,s)ds \in H^2(0,1) \cap H_0^1(0,1), u^7(x,0) = 0 \Big\}. \end{aligned}$$

It is easy to prove that the operator $\mathcal{A}$ is the infinitesimal generator of a C_0 contraction semigroup (see, e.g., [3]).

We shall use the well-known result from the semigroup theory (see, e.g., [19, 155]).

4.3 Polynomial Stability for $\frac{k}{\rho_1} \neq \frac{b}{\rho_2}$

In this section, we shall show that the system is polynomially stable provided that the condition

$$\frac{k}{\rho_1} \neq \frac{b}{\rho_2} \tag{4.3.1}$$

holds.

Our main result is the following estimate of a polynomial-type decay rate.

Theorem 4.3.1 *If $\frac{k}{\rho_1} \neq \frac{b}{\rho_2}$, then there exists a constant $C > 0$, independent of $U_0 \in D(\mathcal{A})$, such that*

$$\| S(t)U_0 \|_{\mathcal{H}} \leq C \left(\frac{\log t}{t}\right)^{\frac{1}{4}} \| U_0 \|_{D(\mathcal{A})} . \tag{4.3.2}$$

We will use Lemma 1.3.23, to prove our main result.

Proof. First we shall show the first condition $i\mathbb{R} \subset \rho(\mathcal{A})$ in (1.3.72) using contradiction arguments. In fact, suppose that it is not true, then there exists a constant $\omega \in \mathbb{R}$, a sequence $\{\beta_n\} \in \mathbb{R}$ with $\beta_n \to \omega,\ |\beta_n| < |\omega|$ and a sequence of functions

$$U_n = \left(u_n^1, u_n^2, u_n^3, u_n^4, u_n^5, u_n^6, u_n^7\right) \in D(\mathcal{A}) \quad \text{with} \quad \| U_n \|_{\mathcal{H}} = 1 \tag{4.3.3}$$

such that, as $n \to +\infty$,

$$i\beta_n U_n - \mathcal{A}U_n \to 0 \ \text{ in } \ \mathcal{H}, \tag{4.3.4}$$

that is,

$$\begin{cases} i\beta_n u_n^1 - u_n^2 \to 0 \ \text{ in } \ H_*^1(0,1), & (4.3.5)\\ i\beta_n \rho_1 u_n^2 - k\left(u_{n,x}^1 + u_n^3\right)_x \to 0 \ \text{ in } \ L_*^2(0,1), & (4.3.6)\\ i\beta_n u_n^3 - u_n^4 \to 0 \ \text{ in } \ H_0^1(0,1), & (4.3.7)\\ i\beta_n \rho_2 u_n^4 - \bar{b} u_{n,xx}^3 - \int_0^{+\infty} g(s) u_{n,xx}^7 ds + k\left(u_{n,x}^1 + u_n^3\right) + \delta u_{n,x}^6 \to 0 \ \text{ in } \ L^2(0,1), & (4.3.8)\\ i\beta_n u_n^5 - u_n^6 \to 0 \ \text{ in } \ H_0^1(0,1), & (4.3.9)\\ i\beta_n \rho_3 u_n^6 - \beta u_{n,xx}^5 - \beta u_{n,xx}^6 + \delta u_{n,x}^4 \to 0 \ \text{ in } \ L^2(0,1), & (4.3.10)\\ i\beta_n u_n^7 + u_{n,s}^7 - u_n^4 \to 0 \ \text{ in } \ L_g^2(\mathbb{R}^+, H_0^1). & (4.3.11) \end{cases}$$

Taking the inner product of (4.3.4) with U_n in $\mathcal{H}$ and then taking its real part, we get

$$-Re < \mathcal{A}U_n, U_n >_{\mathcal{H}} = -\frac{1}{2}\int_0^1 \int_0^{+\infty} g'(s) \mid u_{n,x}^7 \mid^2 dsdx + \beta \int_0^1 \mid u_{n,x}^6 \mid^2 dx \to 0.$$

Using the hypotheses on g, we find that

$$\begin{cases} u_n^7 \to 0 \text{ in } L_g^2(\mathbb{R}^+, H_0^1), & (4.3.12) \\ u_n^6 \to 0 \text{ in } H_0^1(0,1) \hookrightarrow L^2(0,1), & (4.3.13) \end{cases}$$

which, with (4.3.13) and (4.3.9), gives

$$u_n^5 \to 0 \text{ in } H_0^1(0,1) \hookrightarrow L^2(0,1). \tag{4.3.14}$$

Then using (4.3.3), we find that

$$\rho_1 \parallel u_n^2 \parallel_{L^2}^2 + \rho_2 \parallel u_n^4 \parallel_{L^2}^2 + \beta \parallel u_{n,x}^3 \parallel_{L^2}^2 + k \parallel u_{n,x}^1 + u_n^3 \parallel_{L^2}^2 \to 1. \tag{4.3.15}$$

On the other hand, taking the inner product of (4.3.5) with $\rho_1 u_n^2$ in L_*^2 and (4.3.6) with $\rho_1 u_n^1$ in L_*^2, respectively, we obtain

$$i\rho_1\beta_n\left(u_n^1, u_n^2\right)_{L^2} - \rho_1 \parallel u_n^2 \parallel_{L^2}^2 \to 0$$

and

$$i\rho_1\beta_n\left(u_n^2, u_n^1\right)_{L^2} - k\left(u_{n,x}^1 + u_n^3, u_{n,x}^1\right)_{L^2} \to 0.$$

Adding and taking the real part, we get

$$kRe\left(u_{n,x}^1 + u_n^3, u_{n,x}^1\right)_{L^2} - \rho_1 \parallel u_n^2 \parallel_{L^2}^2 \to 0. \tag{4.3.16}$$

Analogously, taking the inner product of (4.3.7) with $\rho_2 u_n^4$ in $L^2(0,1)$ and (4.3.8) with u_n^3 in $L^2(0,1)$, respectively, we have

$$i\rho_2\beta_n\left(u_n^3, u_n^4\right)_{L^2} - \rho_2 \parallel u_n^4 \parallel_{L^2}^2 \to 0 \tag{4.3.17}$$

and

$$i\rho_2\beta_n\left(u_n^4, u_n^3\right)_{L^2} + \bar{b} \parallel u_{n,x}^3 \parallel_{L^2}^2 + \int_0^{+\infty} g(s)\left(u_{n,x}^7, u_{n,x}^3\right)_{L^2} \to 0. \tag{4.3.18}$$

Noting (4.3.12) and (4.3.13), we have

$$\int_0^{+\infty} g(s)\left(u_{n,x}^7, u_{n,x}^3\right)_{L^2} ds + \delta\left(u_{n,x}^6, u_n^3\right)_{L^2} \to 0,$$

which, used in (4.3.18), results in

$$i\rho_2\beta_n\Big(u_n^4, u_n^3\Big)_{L^2} + \bar{b} \parallel u_{n,x}^3 \parallel_{L^2}^2 + k\Big(u_{n,x}^1 + u_n^3, u_n^3\Big)_{L^2} \to 0. \tag{4.3.19}$$

Adding (4.3.17) and (4.3.19) and taking real part, we get

$$-\rho_2 \parallel u_n^4 \parallel_{L^2}^2 + \bar{b} \parallel u_{n,x}^3 \parallel_{L^2}^2 + kRe\Big(u_{n,x}^1 + u_n^3, u_n^3\Big)_{L^2} \to 0. \tag{4.3.20}$$

Adding (4.3.16) with (4.3.20), we have

$$\bar{b} \parallel u_{n,x}^3 \parallel_{L^2}^2 + k \parallel u_{n,x}^1 + u_n^3 \parallel_{L^2}^2 - \rho_1 \parallel u_n^2 \parallel_{L^2}^2 - \rho_2 \parallel u_n^2 \parallel_{L^2}^4 \to 0. \tag{4.3.21}$$

Consequently, from (4.3.15) and (4.3.21), it follows

$$\begin{cases} \bar{b} \parallel u_{n,x}^3 \parallel_{L^2}^2 + k \parallel u_{n,x}^1 + u_n^3 \parallel_{L^2}^2 \to \dfrac{1}{2}, & (4.3.22) \\ \rho_1 \parallel u_n^2 \parallel_{L^2}^2 + \rho_2 \parallel u_n^2 \parallel_{L^2}^4 \to \dfrac{1}{2}. & (4.3.23) \end{cases}$$

Also it is clear that $s \mapsto \frac{1}{\beta_n^2} u_n^4 \in L_g^2(\mathbb{R}^+, H_0^1)$. Then multiplying (4.3.11) with $\frac{1}{\beta_n^2} u_n^4$ in $L_g^2(\mathbb{R}^+, H_0^1)$ gives

$$i\left(u_n^7, \frac{u_n^4}{\beta_n}\right)_{L_g^2} + \frac{1}{\beta_n^2}\left(u_{n,s}^7, u_n^4\right)_{L_g^2} - \frac{1}{\beta_n^2}\left(u_n^4, u_n^4\right)_{L_g^2} \to 0. \tag{4.3.24}$$

Using (4.3.7), we derive that $\frac{u_n^4}{\beta_n}$ is bounded in $H_0^1(0, 1)$, and using (4.3.12), we get that the first term of (4.3.24) converges to zero. This yields

$$b_0 \parallel \frac{u_n^4}{\beta_n} \parallel_{H_0^1}^2 - \frac{1}{\beta_n^2} \int_0^{+\infty} g(s)\Big(u_{n,s}^7, u_n^4\Big)_{H_0^1} ds \to 0, \tag{4.3.25}$$

where $b_0 = \int_0^{+\infty} g(s)ds$. We now prove that the second term in (4.3.25) converges to zero. In fact, using again the fact that $\frac{u_n^4}{\beta_n}$ is bounded in $H_0^1(0, 1)$, (4.2.1) and (4.3.12), we have

$$\begin{aligned} \left| -\frac{1}{\beta_n^2} \int_0^{+\infty} g(s) \left(u_{n,s}^7, u_n^4\right)_{H_0^1} ds \right| &= \frac{1}{|\beta_n|} \left| -\int_0^{+\infty} g'(s)\left(u_{n,s}^7, \frac{u_n^4}{\beta_n}\right)_{H_0^1} ds \right| \\ &\le \frac{k_0}{|\beta_n|} \parallel \frac{u_n^4}{\beta_n} \parallel_{H_0^1} \int_0^{+\infty} g(s) \parallel u_n^7(s) \parallel_{H_0^1} ds \\ &\le \frac{k_0\sqrt{b_0}}{|\beta_n|} \parallel \frac{u_n^4}{\beta_n} \parallel_{H_0^1} \parallel u_n^7 \parallel_{L_g^2} \to 0. \end{aligned} \tag{4.3.26}$$

Therefore it follows from (4.3.25) that

$$\frac{u_n^4}{\beta_n} \to 0 \quad \text{in} \quad H_0^1(0,1),$$

which, together with (4.3.7), implies

$$u_n^3 \to 0 \quad \text{in} \quad H_0^1(0,1). \tag{4.3.27}$$

Using (4.3.27) in (4.3.22), we get

$$k \parallel u_{n,x}^1 + u_n^3 \parallel_{L^2}^2 \to \frac{1}{2}. \tag{4.3.28}$$

We want to show that this is a contradiction if the basic condition (4.3.2) holds. In fact, multiplying (4.3.8) by $(u_{n,x}^1 + u_n^3)$ in $L^2(0,1)$, we have

$$\begin{aligned} i\beta_n\rho_2\Big(u_n^4, u_{n,x}^1 + u_n^3\Big)_{L^2} - \Big(\bar{b}u_{n,xx}^3 + \int_0^{+\infty} g(s)u_{n,x}^7(s)ds, (u_{n,x}^1 + u_n^3)_x\Big)_{L^2} \\ + k \parallel u_{n,x}^1 + u_n^3 \parallel_{L^2}^2 + \delta\Big(u_{n,x}^6, u_{n,x}^1 + u_n^3\Big)_{L^2} \to 0. \end{aligned} \tag{4.3.29}$$

Noting that by (4.3.13), we have that the last term of (4.3.29) converges to zero, we infer

$$\begin{aligned} i\beta_n\rho_2\Big(u_n^4, u_{n,x}^1 + u_n^3\Big)_{L^2} - \Big(\bar{b}u_{n,xx}^3 + \int_0^{+\infty} g(s)u_{n,x}^7(s)ds, (u_{n,x}^1 + u_n^3)_x\Big)_{L^2} \\ + k \parallel u_{n,x}^1 + u_n^3 \parallel_{L^2}^2 \to 0. \end{aligned} \tag{4.3.30}$$

Also multiplying (4.3.6) by $\frac{1}{k}\Big(\bar{b}u_{n,x}^3 + \int_0^{+\infty} g(s)u_{n,x}^7(s)ds\Big)$ in $L^2(0,1)$ results in

$$\begin{aligned} -i\frac{\rho_1}{k}\beta_n\Big(\bar{b}u_{n,x}^3 + \int_0^{+\infty} g(s)u_{n,x}^7(s)ds, u_n^2\Big)_{L^2} \\ - \Big(\bar{b}u_{n,x}^3 + \int_0^{+\infty} g(s)u_{n,x}^7(s)ds, \Big(u_{n,x}^1 + u_n^3\Big)_x\Big)_{L^2} \to 0. \end{aligned} \tag{4.3.31}$$

Then adding (4.3.30) and (4.3.31), we obtain

$$\begin{aligned} i\beta_n\rho_2\Big(u_n^4, u_{n,x}^1 + u_n^3\Big)_{L^2} - i\frac{\rho_1}{k}\beta_n\Big(\bar{b}u_{n,x}^3 + \int_0^{+\infty} g(s)u_{n,x}^7(s)ds, u_n^2\Big)_{L^2} \\ + k \parallel u_{n,x}^1 + u_n^3 \parallel_{L^2}^2 \to 0. \end{aligned} \tag{4.3.32}$$

On the other hand, multiplying (4.3.5) by $\rho_2 u_n^4$, (4.3.7) by $i\rho_2\beta_n u_n^3$ and $-\rho_2 u_{n,x}^2$ in $L^2(0,1)$, respectively, we get

$$-i\beta_n\rho_2\Big(u_n^4, u_{n,x}^1\Big)_{L^2} - \rho_2\Big(u_n^4, u_{n,x}^2\Big)_{L^2} \to 0, \tag{4.3.33}$$

$$\beta_n^2\rho_2 \parallel u_n^3 \parallel_{L^2}^2 + i\beta_n\rho_2\Big(u_n^4, u_n^3\Big)_{L^2} \to 0, \tag{4.3.34}$$

$$-i\beta_n\rho_2\Big(u_n^3, u_{n,x}^2\Big)_{L^2} + \rho_2\Big(u_n^4, u_{n,x}^2\Big)_{L^2} \to 0. \tag{4.3.35}$$

Since $u_n^3 \to 0$ in $H_0^1(0,1) \hookrightarrow L^2(0,1)$, we obtain from (4.3.34) that

$$i\beta_n\rho_2\Big(u_n^4, u_n^3\Big)_{L^2} \to 0. \tag{4.3.36}$$

Adding (4.3.32), (4.3.33) and (4.3.36), we deduce that

$$-i\beta_n\rho_2\Big(u_n^3, u_{n,x}^2\Big)_{L^2} - i\frac{\rho_1}{k}\beta_n\Big(\bar{b}u_{n,x}^3 + \int_0^{+\infty} g(s)u_{n,x}^7(s)ds, u_n^2\Big)_{L^2}$$
$$+k \parallel u_{n,x}^1 + u_n^3 \parallel_{L^2}^2 \to 0. \tag{4.3.37}$$

Now from (4.3.11), we derive

$$i\beta_n u_{n,x}^7 + u_{n,xs}^7 - u_{n,x}^4 \to 0 \quad \text{in} \quad L_g^2(\mathbb{R}^+, L^2).$$

Multiplying by $\frac{\rho_1}{k}u_n^2$ in $L_g^2(\mathbb{R}^+, L^2)$ implies

$$i\beta_n\frac{\rho_1}{k}\Big(u_{n,x}^7, u_n^2\Big)_{L_g^2(\mathbb{R}^+,L^2)} + \frac{\rho_1}{k}\Big(u_{n,xs}^7, u_n^2\Big)_{L_g^2(\mathbb{R}^+,L^2)} - i\beta_n\frac{\rho_1 b_0}{k}\Big(u_{n,x}^4, u_n^2\Big)_{L^2} \to 0. \tag{4.3.38}$$

Using a similar argument used in (4.3.26), we conclude

$$\frac{\rho_1}{k}\Big(u_{n,sx}^7, u_n^2\Big)_{L_g^2(\mathbb{R}^+,L^2)} \to 0,$$

which, together with (4.3.38), gives

$$\frac{\rho_1 b_0}{k}\Big(u_{n,x}^4, u_n^2\Big)_{L^2} \to 0. \tag{4.3.39}$$

Multiplying (4.3.7) by $-\frac{\rho_1 b_0}{k}u_{n,x}^2$ in $L^2(0,1)$ yields

$$i\beta_n\frac{\rho_1 b_0}{k}\Big(u_n^3, u_{n,x}^2\Big)_{L^2} + \frac{\rho_1 b_0}{k}\Big(u_{n,x}^4, u_n^2\Big)_{L^2} \to 0. \tag{4.3.40}$$

Now adding (4.3.39) and (4.3.40), we get

$$i\beta_n \frac{\rho_1 b_0}{k}\Big(u_n^3, u_{n,x}^2\Big)_{L^2} \to 0. \tag{4.3.41}$$

Finally, adding (4.3.37) and (4.3.41), we obtain

$$-i\beta_n\rho_2\Big(u_n^3, u_{n,x}^2\Big)_{L^2} + i\beta_n\frac{\rho_1\bar{b}}{k}\Big(u_n^3, u_{n,x}^2\Big)_{L^2} + i\frac{\rho_1 b_0}{k}\Big(u_n^3, u_{n,x}^2\Big)_{L^2} + k \parallel u_{n,x}^1 + u_n^3 \parallel_{L^2}^2 \to 0,$$

and using that $\bar{b} = b - b_0$, we obtain

$$i\beta_n b\Big(\frac{\rho_1}{k} - \frac{\rho_2}{b}\Big)\Big(u_n^3, u_{n,x}^2\Big)_{L^2} + k \parallel u_{n,x}^1 + u_n^3 \parallel_{L^2}^2 \to 0,$$

which, together with (4.3.2), implies

$$k \parallel u_{n,x}^1 + u_n^3 \parallel_{L^2}^2 \to 0.$$

This contradicts (4.3.28). Thus the first condition $i\mathbb{R} \subset \rho(\mathcal{A})$ is proved.

Next, we shall show the second conditions in Lemma 1.3.72, satisfied by using contradiction arguments. In fact, suppose that it is not true, there exists a sequence $U_n \in \mathcal{A}$, with $\parallel U_n \parallel_{\mathcal{H}} = 1$ and a sequence $\beta_n \in \mathbb{R}$ with $\beta_n \to +\infty$ such that

$$\lim_{n\to+\infty} \frac{1}{\beta_n^j} \parallel (i\beta_n I - \mathcal{A})U_n \parallel_{\mathcal{H}} = 0, \tag{4.3.42}$$

i.e., in $L^2(0,1)$, we have the following convergence

$$\begin{cases} \beta_n^4\left(i\beta_n u_n^1 - u_n^2\right) \to 0 \text{ in } H_*^1(0,1), & (4.3.43)\\ \beta_n^4\Big((i\beta_n\rho_1 u_n^2 - k\big(u_{n,x}^1 + u_n^3\big)_x\Big) \to 0 \text{ in } L_*^2(0,1), & (4.3.44)\\ \beta_n^4\left(i\beta_n u_n^3 - u_n^4\right) \to 0 \text{ in } H_0^1(0,1), & (4.3.45)\\ \beta_n^4\Big(i\beta_n\rho_2 u_n^4 - \bar{b}u_{n,xx}^3 - \int_0^{+\infty} g(s)u_{n,xx}^7 ds + k\big(u_{n,x}^1 + u_n^3\big) + \delta u_{n,x}^6\Big) \to 0 \text{ in } L^2(0,1), & (4.3.46)\\ \beta_n^4\Big(i\beta_n u_n^5 - u_n^6\Big) \to 0 \text{ in } H^1(0,1), & (4.3.47)\\ \beta_n^4\Big(i\beta_n\rho_3 u_n^6 - \bar{\beta}u_{n,xx}^5 - \bar{\beta}u_{n,xx}^6 + \delta u_{n,x}^4\Big) \to 0 \text{ in } L^2(0,1), & (4.3.48)\\ \beta_n^4\left(i\beta_n u_n^7 + u_{n,s}^7 - u_n^4\right) \to 0 \text{ in } L_g^2(\mathbb{R}^+, H_0^1). & (4.3.49) \end{cases}$$

Our goal is to derive $\parallel U_n \parallel_{\mathcal{H}} \to 0$ as a contradiction.
From (4.3.42), taking the inner product of $\beta_n^4(i\beta_n I - \mathcal{A})U_n$ with U_n in $\mathcal{H}$ and then taking its real part, we get

$$\begin{aligned} &Re < \beta_n^4(i\beta_n - \mathcal{A})U_n, U_n >_{\mathcal{H}} \\ &= -\frac{1}{2}\int_0^1\int_0^{+\infty} g'(s) \mid u_{n,x}^7 \mid^2 \beta_n^4 ds dx + \bar{\beta}\int_0^1 \mid u_{n,x}^6 \mid^2 \beta_n^4 dx \to 0. \end{aligned} \tag{4.3.50}$$

Using the hypotheses on g, we find that

$$\begin{cases} \beta_n^{2'} u_n^7 \to 0 \quad \text{in} \quad L_g^2(\mathbb{R}^+, H_0^1), & (4.3.51) \\ \beta_n^2 u_n^6 \to 0 \quad \text{in} \quad H_0^1(0,1) \hookrightarrow L^2(0,1). & (4.3.52) \end{cases}$$

Dividing (4.3.48) by β_n^3, and taking the inner product with $u_{n,x}^6$ in $L^2(0,1)$, then we get

$$i\rho_3 \parallel \beta_n u_n^6 \parallel^2 - \bar{\beta} < \beta_n u_{n,xx}^5, u_n^6 > - \bar{\beta} < \beta_n u_{n,xx}^6, u_n^6 > + \delta < \frac{u_{n,x}^4}{\beta_n}, \beta_n^2 u_n^6 > \to 0, \tag{4.3.53}$$

then, replacing u_n^6 by $i\beta_n u_n^5$ and using (4.3.52), we obtain

$$\parallel \beta_n u_{n,x}^5 \parallel^2 \to 0. \tag{4.3.54}$$

Dividing (4.3.48) by β_n^3 and replacing $\beta_n u_n^4$ by $i\beta_n^2 u_n^3$ in view of (4.3.45), using (4.3.52), we deduce

$$-\bar{\beta}\beta_n u_{n,xx}^5 - \bar{\beta}\beta_n u_{n,xx}^6 + i\delta\beta_n^2 u_{n,x}^3 \to 0 \quad \text{in} \quad L^2(0,1). \tag{4.3.55}$$

Now taking the inner product of the above with $\frac{u_{n,x}^3}{\beta_n^2}$ in $L^2(0,1)$ leads to

$$\begin{aligned} & -\bar{\beta}\beta_n < u_{n,xx}^5, \frac{u_{n,x}^3}{\beta_n^2} > - \bar{\beta}\beta_n < u_{n,xx}^6, \frac{u_{n,x}^3}{\beta_n^2} > + i\delta \parallel u_{n,x}^3 \parallel^2 \\ & = \bar{\beta} < \beta_n u_{n,x}^5, \frac{u_{n,xx}^3}{\beta_n^2} > + \bar{\beta} < \beta_n u_{n,x}^6, \frac{u_{n,xx}^3}{\beta_n^2} > + i\delta \parallel u_{n,x}^3 \parallel^2 \to 0 \text{ in } L^2(0,1), \end{aligned} \tag{4.3.56}$$

which, along with (4.3.52) and (4.3.54), implies that

$$\parallel u_{n,x}^3 \parallel \to 0, \tag{4.3.57}$$

because of (4.3.50) and the uniform boundedness of $\parallel \frac{u_{n,xx}^3}{\beta_n} \parallel$ in n.
Dividing (4.3.48) by β_n^4, taking inner product with $u_{n,x}^4$ in $L^2(0,1)$, and using (4.3.52), (4.3.54), we can obtain

$$\parallel u_{n,x}^4 \parallel \to 0. \tag{4.3.58}$$

Now take the inner product of (4.3.44) with $\frac{u_n^2}{\beta_n^3}$ to get

$$i\rho_1 \parallel \beta_n u_n^2 \parallel^2 -k < u_{n,xx}^1, \beta_n u_n^2 > -k < \beta_n u_{n,x}^3, u_n^2 > \to 0, \tag{4.3.59}$$

which, by using (4.3.43), (4.3.57) and integrating by part, gives

$$i\rho_1 \parallel \beta_n u_n^2 \parallel^2 +ik \parallel \beta_n u_{n,x}^1 \parallel^2 \to 0. \tag{4.3.60}$$

Let $f_n(x) = \int_0^x (u_{n,x}^1 + u_n^3)dx$ and $f_n'(x) = u_{n,x}^1 + u_n^3$, we can rewrite the $L^2(0,1)$ inner product of (4.3.46) and $\frac{f_n'}{\beta_n^4}$ as

$$- < i\rho_2 u_{n,x}^4, \beta_n f_n > +\bar{b} < \beta_n u_{n,x}^3, \frac{f_n''}{\beta_n} > + < \beta_n \int_0^{+\infty} g(s) u_{n,x}^7, \frac{f_n''}{\beta_n} >$$
$$+k \parallel f_n' \parallel^2 +\delta < \beta_n^2 u_{n,x}^6, \frac{f_n'}{\beta_n} > \to 0. \tag{4.3.61}$$

It follows from (4.3.57), (4.3.60) and (4.3.44) that $\parallel \beta_n f_n \parallel$ and $\frac{f_n''}{\beta_n}$ are both uniformly bounded in n, which implies that

$$\parallel f_n' \parallel = \parallel u_{n,x}^1 + u_n^3 \parallel \to 0. \tag{4.3.62}$$

Now every term in $\parallel U_n \parallel_{\mathcal{H}}$ converges to zero, which is a contradiction. Thus, the proof is complete. □

4.4 Exponential Stability for $\frac{k}{\rho_1} = \frac{b}{\rho_2}$

In this section, we shall show that the system is exponentially stable as time tends to $+\infty$ provided that the condition

$$\frac{k}{\rho_1} = \frac{b}{\rho_2} \tag{4.4.1}$$

holds. In order to use Lemma 1.3.22, we have to verify if the following two conditions hold,

$$i\mathbb{R} \subset \rho(\mathcal{A}) \tag{4.4.2}$$

and

$$\exists\, C > 0,\ \forall\, \lambda \in \mathbb{R} : \parallel (i\lambda I - \mathcal{A})^{-1} \parallel_{\mathcal{H}} \leq C. \tag{4.4.3}$$

First we shall show (4.4.2) using contradiction arguments. In fact, suppose that (4.4.2) is not true. Then there exists a constant $\omega \in \mathbb{R}$, a sequence $\{\beta_n\} \subset \mathbb{R}$ with

$\beta_n \to \omega,\ |\beta_n| < |\omega|$ and a sequence of functions

$$U_n = \left(u_n^1, u_n^2, u_n^3, u_n^4, u_n^5, u_n^6, u_n^7\right) \in D(\mathcal{A}) \quad \text{with} \quad \| U_n \|_{\mathcal{H}} = 1 \tag{4.4.4}$$

such that, as $n \to +\infty$,

$$i\beta_n U_n - \mathcal{A}U_n \to 0 \quad \text{in} \quad \mathcal{H}, \tag{4.4.5}$$

that is,

$$\begin{cases} i\beta_n u_n^1 - u_n^2 \to 0 \quad \text{in} \quad H_*^1(0,1), & (4.4.6)\\ i\beta_n\rho_1 u_n^2 - k\left(u_{n,x}^1 + u_n^3\right)_x \to 0 \quad \text{in} \quad L_*^2(0,1), & (4.4.7)\\ i\beta_n u_n^3 - u_n^4 \to 0 \quad \text{in} \quad H_0^1(0,1), & (4.4.8)\\ i\beta_n\rho_2 u_n^4 - \bar{b}u_{n,xx}^3 - \int_0^{+\infty} g(s)u_{n,xx}^7 ds + k\left(u_{n,x}^1 + u_n^3\right) + \delta u_{n,x}^6 \to 0 \quad \text{in} \quad L^2(0,1), & (4.4.9)\\ i\beta_n u_n^5 - u_n^6 \to 0 \quad \text{in} \quad H_0^1(0,1), & (4.4.10)\\ i\beta_n\rho_3 u_n^6 - \beta u_{n,xx}^5 - \beta u_{n,xx}^6 + \delta u_{n,x}^4 \to 0 \quad \text{in} \quad L^2(0,1), & (4.4.11)\\ i\beta_n u_n^7 + u_{n,s}^7 - u_n^4 \to 0 \quad \text{in} \quad L_g^2(\mathbb{R}^+, H_0^1). & (4.4.12) \end{cases}$$

Taking the inner product of (4.4.5) with U_n in $\mathcal{H}$ and then taking its real part, we get

$$-Re\langle \mathcal{A}U_n, U_n\rangle_{\mathcal{H}} = -\frac{1}{2}\int_0^1\int_0^{+\infty} g'(s) \mid u_{n,x}^7 \mid^2 dsdx + \beta\int_0^1 \mid u_{n,x}^6 \mid^2 dx \to 0.$$

Using the hypotheses on g, we find that

$$\begin{cases} u_n^7 \to 0 \quad \text{in} \quad L_g^2(\mathbb{R}^+, H_0^1), & (4.4.13)\\ u_n^6 \to 0 \quad \text{in} \quad H_0^1(0,1) \hookrightarrow L^2(0,1), & (4.4.14) \end{cases}$$

which, by inserting (4.4.14) into(3.4.10), implies

$$u_n^5 \to 0 \quad \text{in} \quad H_0^1(0,1) \hookrightarrow L^2(0,1). \tag{4.4.15}$$

Then using (4.4.4), we find that

$$\rho_1 \| u_n^2 \|_{L^2}^2 + \rho_2 \| u_n^4 \|_{L^2}^2 + \beta \| u_{n,x}^3 \|_{L^2}^2 + k \| u_{n,x}^1 + u_n^3 \|_{L^2}^2 \to 1. \tag{4.4.16}$$

On the other hand, denoting $(f, g)_I$ to represent the inner product of f, g in the space I, taking the inner product of (4.4.6) with $\rho_1 u_n^2$ in L_*^2 and (4.4.7) with $\rho_1 u_n^1$ in L_*^2, respectively, we have

$$i\rho_1\beta_n\left(u_n^1, u_n^2\right)_{L^2} - \rho_1 \| u_n^2 \|_{L^2}^2 \to 0$$

and

$$i\rho_1\beta_n\Big(u_n^2, u_n^1\Big)_{L^2} - k\Big(u_{n,x}^1 + u_n^3, u_{n,x}^1\Big)_{L^2} \to 0.$$

Adding and taking the real part of the above two relations, we get

$$kRe\Big(u_{n,x}^1 + u_n^3, u_{n,x}^1\Big)_{L^2} - \rho_1 \parallel u_n^2 \parallel_{L^2}^2 \to 0. \tag{4.4.17}$$

Analogously, taking the inner product of (4.4.8) with $\rho_2 u_n^4$ in $L^2(0, 1)$ and (4.4.9) with u_n^3 in $L^2(0, 1)$, respectively, we derive

$$i\rho_2\beta_n\Big(u_n^3, u_n^4\Big)_{L^2} - \rho_2 \parallel u_n^4 \parallel_{L^2}^2 \to 0 \tag{4.4.18}$$

and

$$i\rho_2\beta_n\Big(u_n^4, u_n^3\Big)_{L^2} + \bar{b} \parallel u_{n,x}^3 \parallel_{L^2}^2 + \int_0^{+\infty} g(s)\Big(u_{n,x}^7, u_{n,x}^3\Big)_{L^2} \to 0. \tag{4.4.19}$$

Noting (4.4.13) and (4.4.14), we have

$$\int_0^{+\infty} g(s)\Big(u_{n,x}^7, u_{n,x}^3\Big)_{L^2} ds + \delta\Big(u_{n,x}^6, u_n^3\Big)_{L^2} \to 0,$$

which, used in (4.4.19), results in

$$i\rho_2\beta_n\Big(u_n^4, u_n^3\Big)_{L^2} + \bar{b} \parallel u_{n,x}^3 \parallel_{L^2}^2 + k\Big(u_{n,x}^1 + u_n^3, u_n^3\Big)_{L^2} \to 0. \tag{4.4.20}$$

Adding (4.4.18) and (4.4.20) and taking real part, we get

$$-\rho_2 \parallel u_n^4 \parallel_{L^2}^2 + \bar{b} \parallel u_{n,x}^3 \parallel_{L^2}^2 + kRe\Big(u_{n,x}^1 + u_n^3, u_n^3\Big)_{L^2} \to 0. \tag{4.4.21}$$

Adding (4.4.17) with (4.4.21), we have

$$\bar{b} \parallel u_{n,x}^3 \parallel_{L^2}^2 + k \parallel u_{n,x}^1 + u_n^3 \parallel_{L^2}^2 - \rho_1 \parallel u_n^2 \parallel_{L^2}^2 - \rho_2 \parallel u_n^2 \parallel_{L^2}^4 \to 0. \tag{4.4.22}$$

Consequently, from (4.4.16) and (4.4.22), it follows that

$$\begin{cases} \bar{b} \parallel u_{n,x}^3 \parallel_{L^2}^2 + k \parallel u_{n,x}^1 + u_n^3 \parallel_{L^2}^2 \to \dfrac{1}{2}, & (4.4.23) \\ \rho_1 \parallel u_n^2 \parallel_{L^2}^2 + \rho_2 \parallel u_n^2 \parallel_{L^2}^4 \to \dfrac{1}{2}. & (4.4.24) \end{cases}$$

Also it is clear that $s \mapsto \frac{1}{\beta_n^2} u_n^4 \in L_g^2(\mathbb{R}^+, H_0^1)$. Then multiplying (4.4.12) with $\frac{1}{\beta_n^2} u_n^4$ in $L_g^2(\mathbb{R}^+, H_0^1)$ gives

$$i\Big(u_n^7, \frac{u_n^4}{\beta_n}\Big)_{L_g^2} + \frac{1}{\beta_n^2}\Big(u_{n,s}^7, u_n^4\Big)_{L_g^2} - \frac{1}{\beta_n^2}\Big(u_n^4, u_n^4\Big)_{L_g^2} \to 0. \tag{4.4.25}$$

Using (4.4.8), we have that $\frac{u_n^4}{\beta_n}$ is bounded in $H_0^1(0,1)$, and using (4.4.13), we get that the first term of (4.4.26) converges to zero. This yields

$$b_0 \parallel \frac{u_n^4}{\beta_n} \parallel_{H_0^1}^2 - \frac{1}{\beta_n^2}\int_0^{+\infty} g(s)\Big(u_{n,s}^7, u_n^4\Big)_{H_0^1} ds \to 0, \tag{4.4.26}$$

where $b_0 = \int_0^{+\infty} g(s)ds$. We now prove that the second term in (4.4.26) converges to zero. In fact, using again the fact that $\frac{u_n^4}{\beta_n}$ is bounded in $H_0^1(0,1)$, (4.2.1) and (4.4.13), we have

$$\begin{aligned}
\left|-\frac{1}{\beta_n^2}\int_0^{+\infty} g(s)\Big(u_{n,s}^7, u_n^4\Big)_{H_0^1} ds\right| &= \frac{1}{|\beta_n|}\left|-\int_0^{+\infty} g'(s)\Big(u_{n,s}^7, \frac{u_n^4}{\beta_n}\Big)_{H_0^1} ds\right| \\
&\le \frac{k_0}{|\beta_n|} \parallel \frac{u_n^4}{\beta_n} \parallel_{H_0^1} \int_0^{+\infty} g(s) \parallel u_n^7(s) \parallel_{H_0^1} ds \\
&\le \frac{k_0\sqrt{b_0}}{|\beta_n|} \parallel \frac{u_n^4}{\beta_n} \parallel_{H_0^1} \parallel u_n^7 \parallel_{L_g^2} \to 0.
\end{aligned} \tag{4.4.27}$$

Therefore we can derive from (4.4.26) that

$$\frac{u_n^4}{\beta_n} \to 0 \quad \text{in} \quad H_0^1(0,1),$$

which, along with (4.4.8), implies

$$u_n^3 \to 0 \quad \text{in} \quad H_0^1(0,1), \tag{4.4.28}$$

and using (4.4.28) in (4.4.23), we get

$$k \parallel u_{n,x}^1 + u_n^3 \parallel_{L^2}^2 \to \frac{1}{2}. \tag{4.4.29}$$

We want to show that this is a contradiction if the basic condition (4.4.1) holds. Multiplying (4.4.9) by $(u_{n,x}^1 + u_n^3)$ in $L^2(0,1)$, we have

$$\begin{aligned}
&i\beta_n\rho_2\Big(u_n^4, u_{n,x}^1 + u_n^3\Big)_{L^2} - \Big(\bar{b}u_{n,xx}^3 + \int_0^{+\infty} g(s)u_{n,x}^7(s)ds, (u_{n,x}^1 + u_n^3)_x\Big)_{L^2} \\
&+k \parallel u_{n,x}^1 + u_n^3 \parallel_{L^2}^2 + \delta\Big(u_{n,x}^6, u_{n,x}^1 + u_n^3\Big)_{L^2} \to 0.
\end{aligned} \tag{4.4.30}$$

Noting (4.4.14), we know that the last term of (4.4.30) converges to zero. Then we get

$$i\beta_n\rho_2\Big(u_n^4, u_{n,x}^1 + u_n^3\Big)_{L^2} - \Big(\bar{b}u_{n,xx}^3 + \int_0^{+\infty} g(s)u_{n,x}^7(s)ds, (u_{n,x}^1 + u_n^3)_x\Big)_{L^2}$$
$$+k \parallel u_{n,x}^1 + u_n^3 \parallel_{L^2}^2 \to 0. \tag{4.4.31}$$

Also multiplying (4.4.7) by $\frac{1}{k}\Big(\bar{b}u_{n,x}^3 + \int_0^{+\infty} g(s)u_{n,x}^7(s)ds\Big)$ in $L^2(0,1)$ results in

$$-i\frac{\rho_1}{k}\beta_n\Big(\bar{b}u_{n,x}^3 + \int_0^{+\infty} g(s)u_{n,x}^7(s)ds, u_n^2\Big)_{L^2}$$
$$-\Big(\bar{b}u_{n,x}^3 + \int_0^{+\infty} g(s)u_{n,x}^7(s)ds, \Big(u_{n,x}^1 + u_n^3\Big)_x\Big)_{L^2} \to 0. \tag{4.4.32}$$

Then adding (4.4.31) and (4.4.32), we obtain

$$i\beta_n\rho_2\Big(u_n^4, u_{n,x}^1 + u_n^3\Big)_{L^2} - i\frac{\rho_1}{k}\beta_n\Big(\bar{b}u_{n,x}^3 + \int_0^{+\infty} g(s)u_{n,x}^7(s)ds, u_n^2\Big)_{L^2}$$
$$+k \parallel u_{n,x}^1 + u_n^3 \parallel_{L^2}^2 \to 0. \tag{4.4.33}$$

On the other hand, multiplying (4.4.6) by $\rho_2 u_n^4$, (4.4.8) by $i\rho_2\beta_n u_n^3$ and $-\rho_2 u_{n,x}^2$ in $L^2(0,1)$, respectively, we get

$$\begin{cases} -i\beta_n\rho_2\Big(u_n^4, u_{n,x}^1\Big)_{L^2} - \rho_2\Big(u_n^4, u_{n,x}^2\Big)_{L^2} \to 0, & (4.4.34)\\ \beta_n^2\rho_2 \parallel u_n^3 \parallel_{L^2}^2 + i\beta_n\rho_2\Big(u_n^4, u_n^3\Big)_{L^2} \to 0, & (4.4.35)\\ -i\beta_n\rho_2\Big(u_n^3, u_{n,x}^2\Big)_{L^2} + \rho_2\Big(u_n^4, u_{n,x}^2\Big)_{L^2} \to 0. & (4.4.36)\end{cases}$$

Since $u_n^3 \to 0$ in $H_0^1(0,1) \hookrightarrow L^2(0,1)$, we derive from (4.4.35) that

$$i\beta_n\rho_2\Big(u_n^4, u_n^3\Big)_{L^2} \to 0. \tag{4.4.37}$$

Adding (4.4.33), (4.4.34) and (4.4.37), we deduce that

$$-i\beta_n\rho_2\Big(u_n^3, u_{n,x}^2\Big)_{L^2} - i\frac{\rho_1}{k}\beta_n\Big(\bar{b}u_{n,x}^3 + \int_0^{+\infty} g(s)u_{n,x}^7(s)ds, u_n^2\Big)_{L^2}$$
$$+k \parallel u_{n,x}^1 + u_n^3 \parallel_{L^2}^2 \to 0. \tag{4.4.38}$$

Now from (4.4.12), we have

$$i\beta_n u_{n,x}^7 + u_{n,xs}^7 - u_{n,x}^4 \to 0 \quad \text{in} \quad L_g^2(\mathbb{R}^+, L^2),$$

then multiplying by $\frac{\rho_1}{k}u_n^2$ in $L_g^2(\mathbb{R}^+, L^2)$ results in

$$i\beta_n\frac{\rho_1}{k}\left(u_{n,x}^7, u_n^2\right)_{L_g^2(\mathbb{R}^+,L^2)} + \frac{\rho_1}{k}\left(u_{n,xs}^7, u_n^2\right)_{L_g^2(\mathbb{R}^+,L^2)} - i\beta_n\frac{\rho_1 b_0}{k}\left(u_{n,x}^4, u_n^2\right)_{L^2} \to 0. \tag{4.4.39}$$

Using a similar argument used in (4.4.27), we can conclude

$$\frac{\rho_1}{k}\left(u_{n,sx}^7, u_n^2\right)_{L_g^2(\mathbb{R}^+,L^2)} \to 0,$$

which, together with (4.4.39), implies

$$\frac{\rho_1 b_0}{k}\left(u_{n,x}^4, u_n^2\right)_{L^2} \to 0. \tag{4.4.40}$$

Multiplying (4.4.8) by $-\frac{\rho_1 b_0}{k}u_{n,x}^2$ in $L^2(0, 1)$ yields

$$i\beta_n\frac{\rho_1 b_0}{k}\left(u_n^3, u_{n,x}^2\right)_{L^2} + \frac{\rho_1 b_0}{k}\left(u_{n,x}^4, u_n^2\right)_{L^2} \to 0. \tag{4.4.41}$$

Then adding (4.4.40) and (4.4.41), we get

$$i\beta_n\frac{\rho_1 b_0}{k}\left(u_n^3, u_{n,x}^2\right)_{L^2} \to 0. \tag{4.4.42}$$

Finally, adding (4.4.38) and (4.4.42), we obtain

$$-i\beta_n\rho_2\left(u_n^3, u_{n,x}^2\right)_{L^2} + i\beta_n\frac{\rho_1\bar{b}}{k}\left(u_n^3, u_{n,x}^2\right)_{L^2} + i\frac{\rho_1 b_0}{k}\left(u_n^3, u_{n,x}^2\right)_{L^2} + k \parallel u_{n,x}^1 + u_n^3 \parallel_{L^2}^2 \to 0,$$

and using that $\bar{b} = b - b_0$, we obtain

$$i\beta_n b\left(\frac{\rho_1}{k} - \frac{\rho_2}{b}\right)\left(u_n^3, u_{n,x}^2\right)_{L^2} + k \parallel u_{n,x}^1 + u_n^3 \parallel_{L^2}^2 \to 0,$$

and using (4.4.1), we have

$$k \parallel u_{n,x}^1 + u_n^3 \parallel_{L^2}^2 \to 0,$$

which contradicts (4.4.29). Thus (4.4.2) is proved.

To complete the result on exponential stability, we now prove (4.4.3). Note again the resolvent equation $(i\lambda I - \mathcal{A})U = F \in \mathcal{H}$ is given by

$$\begin{cases} i\lambda u^1 - u^2 = f^1, & (4.4.43)\\ i\lambda\rho_1 u^2 - k(u_x^1 + u^3)_x = \rho_1 f^2, & (4.4.44)\\ i\lambda u^3 - u^4 = f^3, & (4.4.45)\\ i\lambda\rho_2 u^4 - \bar{b}u_{xx}^3 - \int_0^{+\infty} g(s)u_{xx}^7(s)ds + k\Big(u_x^1 + u^3\Big) + \delta u_x^6 = \rho_2 f^4, & (4.4.46)\\ i\lambda u^5 - u^6 = f^5, & (4.4.47)\\ i\lambda\rho_3 u^6 - \beta u_{xx}^6 - \beta u_{xx}^5 + \delta u_x^4 = \rho_3 f^6, & (4.4.48)\\ i\lambda u^7 + u_s^7 - u^4 = f^7, & (4.4.49) \end{cases}$$

where $b_0 = \int_0^{+\infty} g(s)ds$, $\bar{b} = b - b_0 > 0$. To prove (4.4.3), we shall use a series of lemmas in the following.

Lemma 4.4.1 *Suppose that conditions (4.2.1) and (4.2.2) on g hold. Then there exists a positive constant C, independent of F, such that*

$$\beta \int_0^1 |u_x^6|^2 dx + \int_0^1 \int_0^{+\infty} g(s)|u_x^7| ds dx \le C \parallel U \parallel_{\mathcal{H}} \parallel F \parallel_{\mathcal{H}} .$$

Proof. Multiplying (4.4.44) by u^2 in $L^2(0, 1)$, we get

$$i\lambda\rho_2 \int_0^1 |u^2|^2 dx + k \int_0^1 (u_x^1 + u^3)\overline{u_x^2} dx = \rho_1 \int_0^1 f^2 \overline{u^2} dx$$

and, using Equation (4.4.43),

$$i\lambda\rho_2 \int_0^1 |u^2|^2 dx - i\lambda k \int_0^1 (u_x^1 + u^3)\overline{u_x^1} dx = \rho_1 \int_0^1 f^2 \overline{u^2} dx + k \int_0^1 (u_x^1 + u^3)\overline{f_x^1} dx. \tag{4.4.50}$$

On the other hand, multiplying equation (4.4.46) by u^4 and integrating over [0, 1], we get

$$\begin{aligned} i\lambda\rho_2 \int_0^1 |u^4|^2 dx + \bar{b} \int_0^1 u_x^3 \overline{u_x^4} dx + \int_0^1 \int_0^{+\infty} g(s) u_x^7 \overline{u_x^4} ds dx \\ + k \int_0^1 \Big(u_x^1 + u^3\Big)\overline{u^4} dx + \delta \int_0^1 u_x^6 \overline{u^4} = \rho_2 \int_0^1 f^4 \overline{u^4} dx. \end{aligned} \tag{4.4.51}$$

Substituting u^4 given by (4.4.49), (4.4.45) into (4.4.51), we get

$$\begin{aligned}
&i\lambda\rho_2\int_0^1 |u^4|^2dx - i\lambda\bar{b}\int_0^1 |u_x^3|^2dx - i\lambda\int_0^1\int_0^{+\infty} g(s)|u_x^7|^2dsdx \\
&\quad - i\lambda k\int_0^1 \left(u_x^1+u^3\right)\overline{u^3}dx + \int_0^1\int_0^{+\infty} g(s)u_x^7\overline{u_{xs}^7}dsdx + \delta\int_0^1 u_x^6\overline{u^4} \\
&= \rho_2\int_0^1 f^4\overline{u^4}dx + \bar{b}\int_0^1 u_x^3\overline{f_x^3}dx + k\int_0^1 \left(u_x^1+u^3\right)\overline{f^3}dx + \int_0^1\int_0^{+\infty} g(s)u_x^7\overline{f_x^7}dsdx.
\end{aligned} \tag{4.4.52}$$

Also, multiplying equation (4.4.48) by $\overline{u^6}$, we obtain

$$i\lambda\rho_3\int_0^1 |u^6|^2dx + \beta\int_0^1 |u_x^6|^2dx - \beta\int_0^1 u_{xx}^5u^6dx = \rho_3\int_0^1 f^6\overline{u^6}dx. \tag{4.4.53}$$

Inserting (4.4.47) into (4.4.53), adding (4.4.51)–(4.4.52), using (4.2.1) and taking the real part, our conclusion follows. □

Lemma 4.4.2 *Under the same hypotheses as in Lemma 4.4.1, there exists a constant $C > 0$ such that*

$$\rho_2\int_0^1 |u^4|^2dx \le C \parallel U \parallel_{\mathcal{H}}\parallel F \parallel_{\mathcal{H}} + C \parallel U \parallel_{\mathcal{H}}^{1/2}\parallel F \parallel_{\mathcal{H}}^{1/2}\left(\parallel u_x^3 \parallel_{L^2} + \parallel u_x^1+u^3 \parallel_{L^2}\right).$$

Proof. Multiplying (4.4.46) by $\int_0^{+\infty} g(s)\overline{u^7}ds$ in $L^2(0,1)$, we get

$$\begin{aligned}
&i\lambda\rho_2\int_0^1\int_0^{+\infty} g(s)\overline{u_x^7}u_x^3dsdx + \int_0^1\left|\int_0^{+\infty} g(s)u_x^7ds\right|^2dx \\
&\quad + k\int_0^1\int_0^{+\infty} g(s)(u_x^1+u^3)\overline{u^7}dsdx - \delta\int_0^1\int_0^{+\infty} g(s)\overline{u_x^7}u^6dsdx \\
&= \rho_2\int_0^1\int_0^{+\infty} g(s)\overline{u^7}f^4dsdx.
\end{aligned} \tag{4.4.54}$$

From Lemma 4.4.1, we obtain

$$\int_0^1\left|\int_0^{+\infty} g(s)u_x^7ds\right|^2dx \le \int_0^{+\infty} g(s)ds\int_0^1\int_0^{+\infty} g(s)|u_x^7|^2dsdx \le C \parallel U \parallel_{\mathcal{H}}\parallel F \parallel_{\mathcal{H}}$$

and

$$\mathrm{Re}\left\{\delta\int_0^1\int_0^{+\infty} g(s)\overline{u_x^7}u^6dsdx\right\} \le C \parallel U \parallel_{\mathcal{H}}\parallel F \parallel_{\mathcal{H}}.$$

Substituting $i\lambda u^7$ given by (4.4.49) into (4.4.54), using

$$\text{Re}\Big\{\delta \int_0^1 \int_0^{+\infty} g(s)\overline{u_s^7} u^4 dsdx\Big\} \le \frac{\rho_2}{2} \int_0^1 |u^4|^2 dx + C \int_0^1 \int_0^{+\infty} |g'(s)||u_x^7|^2 dsdx$$

and using (4.2.1), our conclusion now immediately follows from Lemma 4.4.1. □

Lemma 4.4.3 *Under the same hypotheses as in Lemma 4.4.1, for any $\varepsilon_1 > 0$ there exists a constant $C_{\varepsilon_1} > 0$, at most depending on ε_1, such that*

$$\bar{b} \int_0^1 |\ u_x^3\ |^2\ dx \le C_{\varepsilon_1} \parallel U \parallel_{\mathcal{H}} \parallel F \parallel_{\mathcal{H}}$$
$$+C \parallel U \parallel_{\mathcal{H}}^{1/2} \parallel F \parallel_{\mathcal{H}}^{1/2} \parallel u_x^1 + u^3 \parallel_{L^2} + \varepsilon_1 \rho_1 \parallel u^2 \parallel_{L^2}^2 .$$

Proof. Multiplying (4.4.46) by u^3 yields

$$i\lambda\rho_2 \int_0^1 u^4 \overline{u^3} dx + \bar{b} \int_0^1 |u_x^3|^2 dx + \int_0^1 \int_0^{+\infty} g(s) u_x^7 \overline{u_x^3} dsdx$$
$$+k \int_0^1 (u_x^1 + u^3)\overline{u^3} dx + \delta \int_0^1 u_x^6 \overline{u^3} dx = \rho_2 \int_0^1 f^4 \overline{u^3} dx. \quad (4.4.55)$$

Substituting $i\lambda u^3$ given by (4.4.45) into (4.4.55), we get

$$\bar{b} \int_0^1 |\ u_x^3\ |^2\ dx + k \int_0^1 (u_x^1 + u^3)\overline{u^3} dx = \rho_2 \int_0^1 |u^4|^2 dx$$
$$- \int_0^1 \int_0^{+\infty} g(s) u_x^7 \overline{u_x^3} dsdx + \delta \int_0^1 u^6 \overline{u_x^3} dx + \rho_2 \int_0^1 f^4 \overline{u^3} dx + \rho_2 \int_0^1 u^4 \overline{f^3} dx.$$
$$(4.4.56)$$

On the other hand, multiplying (4.4.44) by $\int_0^x \overline{u^3(y)} dy$, we get

$$i\lambda\rho_1 \int_0^1 u^2 \Big(\int_0^x \overline{u^3(y)} dy\Big) dx - k \int_0^1 (u_x^1 + u^3)_x \Big(\int_0^x \overline{u^3(y)} dy\Big) dx$$
$$= \rho_1 \int_0^1 f^2 \Big(\int_0^x \overline{u^3(y)} dy\Big) dx. \quad (4.4.57)$$

Using (4.4.45), we have

$$k \int_0^1 (u_x^1 + u^3)\overline{u^3} dx = \rho_1 \int_0^1 u^2 \Big(\int_0^x \overline{u^4(y)} dy\Big) dx$$
$$+\rho_1 \int_0^1 u^2 \Big(\int_0^x \overline{f^3(y)} dy\Big) dx + \rho_1 \int_0^1 f^2 \Big(\int_0^x \overline{u^3(y)} dy\Big) dx. \quad (4.4.58)$$

Finally, using (4.4.58) into (4.4.56) and using the estimate

$$\mathrm{Re}\left\{\rho_1\int_0^1 f^2\overline{\int_0^x u^4(y)dydx}\right\}\le \varepsilon_1\rho_1\parallel u^2\parallel_{L^2}^2+C_{\varepsilon_1}\rho_2\parallel u^4\parallel_{L^2}^2,$$

and taking the real part and using Lemmas 4.4.1 and 4.4.2, our conclusion follows finally. □

Our next step is to estimate the term $\parallel u_x^1+u^3\parallel_{L^2}^2$. Here we shall use condition (4.4.1).

Lemma 4.4.4 *Under the same hypotheses as in Lemma 4.4.1 and condition (4.4.1), for any $\varepsilon_2>0$, there exists a constant $C_{\varepsilon_2}>0$, at most depending on ε_2, such that*

$$k\int_0^1 \mid u_x^1+u^3\mid^2 dx\le C_{\varepsilon_2}\parallel U\parallel_{\mathcal{H}}\parallel F\parallel_{\mathcal{H}}+(\varepsilon_1+\varepsilon_2)\rho_1\parallel u^2\parallel_{L^2}^2 .$$

Proof. Multiplying (4.4.46) by $u_x^1+u^3$, we have

$$\begin{aligned}&i\lambda\rho_2\int_0^1 u^4\overline{(u_x^1+u^3)}dx+k\int_0^1\mid u_x^1+u^3\mid dx+\delta\int_0^1 u_x^6\overline{(u_x^1+u^3)}dx\\&+\int_0^1\left(\bar{b}u_x^3+\int_0^{+\infty}g(s)u_x^7ds\right)\overline{(u_x^1+u^3)}_xdx=\rho_2\int_0^1 f^4\overline{(u_x^1+u^3)}dx.\end{aligned}\tag{4.4.59}$$

Substituting $(u_x^1+u^3)_x$ given by (4.4.44) into (4.4.59), we get

$$\begin{aligned}&\underbrace{i\lambda\rho_2\int_0^1 u^4\overline{u_x^1}dx}_{I_1}+\underbrace{i\lambda\rho_2\int_0^1 u^4\overline{u^3}dx}_{I_2}-i\lambda\frac{\bar{b}\rho_1}{k}\int_0^1 u_x^3\overline{u^2}dx+k\int_0^1\mid u_x^1+u^3\mid^2dx\\&+\delta\int_0^1 u_x^6\overline{(u_x^1+u^3)}dx-\underbrace{i\lambda\frac{\rho_1}{k}\int_0^1\int_0^{+\infty}g(s)u_x^7\overline{u^2}dsdx}_{I_3}\\&-\frac{\rho_1}{k}\int_0^1\int_0^{+\infty}g(s)u_x^7\overline{f^2}dsdx-\frac{\bar{b}\rho_1}{k}\int_0^1 u_x^3\overline{f^2}dx\\&=\rho_2\int_0^1 f^4\overline{(u_x^1+u^3)}dx.\end{aligned}\tag{4.4.60}$$

Substituting u^1 given by (4.4.43) and u^4 given by (4.4.45) into I_1, then

$$I_1=i\lambda\rho_2\int_0^1 u^3\overline{u_x^2}dx-\rho_2\int_0^1 u^4\overline{f_x^1}dx+\rho_2\int_0^1 f^3\overline{u_x^2}dx.\tag{4.4.61}$$

Using (4.4.45), we get

$$I_2 = -\rho_2 \int_0^1 | u^4 |^2 \, dx - \rho_2 \int_0^1 u^4 \overline{f^3} dx. \tag{4.4.62}$$

Finally, a substitution of u^7 given by (4.4.49) yields

$$I_3 = \frac{\rho_1}{k} \int_0^1 \int_0^{+\infty} g(s) u^7_{xs} \overline{u^2} ds dx - \frac{\rho_1 b_0}{k} \int_0^1 u^4_x \overline{u^2} dx$$
$$- \frac{\rho_1}{k} \int_0^1 \int_0^{+\infty} g(s) f^7_x \overline{u^2} ds dx.$$

From (4.4.45), we can rewrite I_3 as

$$I_3 = -\frac{\rho_1}{k} \int_0^1 \int_0^{+\infty} g'(s) u^7_x \overline{u^2} ds dx - i\lambda \frac{\rho_1 b_0}{k} \int_0^1 u^3_x \overline{u^2} dx$$
$$+ \frac{\rho_1 b_0}{k} \int_0^1 f^3_x \overline{u^2} dx - \frac{\rho_1}{k} \int_0^1 \int_0^{+\infty} g(s) f^7_x \overline{u^2} ds dx. \tag{4.4.63}$$

Using (4.4.61)–(4.4.63) in (4.4.60), we get

$$i\lambda b \left(\frac{\rho_1}{k} - \frac{\rho_2}{b} \right) \int_0^1 u^3 \overline{u^2_x} dx + k \int_0^1 | u^1_x + u^3 |^2 \, dx$$
$$= \rho_2 \int_0^1 | u^4 |^2 \, dx - \delta \int_0^1 u^6_x \overline{(u^1_x + u^3)} dx$$
$$+ \frac{\rho_1}{k} \int_0^1 \int_0^{+\infty} g'(s) u^7_x \overline{u^2} ds dx + \frac{\rho_1 \bar{b}}{k} \int_0^1 u^3_x \overline{f^2} dx$$
$$+ \frac{\rho_1}{k} \int_0^1 \int_0^{+\infty} g(s) u^7_x \overline{f^2} ds dx + \rho_2 \int_0^1 f^4 \overline{(u^1_x + u^3)} dx$$
$$+ \rho_2 \int_0^1 u^4 \overline{f^3} dx + \rho_2 \int_0^1 u^3 \overline{f^1_x} dx + \left(\rho_2 - \frac{\rho_1 b_0}{k} \right) \int_0^1 f^3_x \overline{u^2} dx$$
$$+ \frac{\rho_1}{k} \int_0^1 \int_0^{+\infty} g(s) f^7_x \overline{u^2} ds dx.$$

Now, using (4.4.1) and the previous lemmas, our claim follows. □

Lemma 4.4.5 *There exists a constant $C > 0$ such that*

$$\rho_1 \int_0^1 | u^2 |^2 \, dx \leq C \parallel U \parallel_{\mathcal{H}} \parallel F \parallel_{\mathcal{H}} + 4k \parallel u^1_x + u^3 \parallel^2_{L^2} .$$

Proof. Multiplying equation (4.4.44) by u^1, we get

$$\underbrace{i\lambda\rho_1\int_0^1 u^2\overline{u^1}dx}_{I_4}+k\int_0^1(u_x^1+u^3)\overline{u_x^1}dx=\rho_1\int_0^1 f^2\overline{u^1}dx.$$

Substituting u^1 given by (4.4.43) into I_4 and taking real parts, we get

$$\rho_1\int_0^1|u^2|^2dx\leq C\parallel U\parallel_{\mathcal{H}}\parallel F\parallel_{\mathcal{H}}+2k\parallel u_x^1+u^3\parallel_{L^2}^2+C\parallel u_x^3\parallel_{L^2}^2.$$

Using Lemmas 4.4.3, and taking ε_1 sufficiently small, our conclusion follows. □

Lemma 4.4.6 *Under the same hypotheses as in Lemma 4.4.1, for any $\varepsilon_3>0$, there exists a constant $C_{\varepsilon_3}>0$, at most depending on ε_3, such that*

$$\frac{\beta}{4}\int_0^1|u_x^5|^2dx\leq C_{\varepsilon_3}\parallel U\parallel_{\mathcal{H}}\parallel F\parallel_{\mathcal{H}}+\frac{2\delta^2}{\beta}\parallel u_x^4\parallel^2 dx. \tag{4.4.64}$$

Proof. Multiplying equation (4.4.48) by u^5, we have

$$i\lambda\rho_3\int_0^1 u^6\overline{u^5}dx-\beta\int_0^1 u_{xx}^6\overline{u^5}dx-\beta\int_0^1 u_{xx}^5\overline{u^5}dx+\delta\int_0^1 u_x^4\overline{u^5}dx=\rho_3\int_0^1 f^6\overline{u^5}dx. \tag{4.4.65}$$

Substituting of u^5 given by (4.4.43) into (4.4.64), taking real parts, and using Lemmas 4.4.1–4.4.5, Young's inequality, our conclusion follows. □

Now we are in the position to prove the main result of this subsection.

Theorem 4.4.1 *Assume hypotheses (4.2.1) and (4.2.2) on g and condition (4.4.1) also hold. Then the heat conduction Timoshenko system (4.1.1)–(4.1.3) is exponentially stable.*

Proof. It suffices to show (4.4.3). Let $U=(u^1,u^2,u^3,u^4,u^5,u^6,u^7)^T$, $F=(f^1,f^2,f^3,f^4,f^5,f^6,f^7)^T$ satisfy (4.4.43)–(4.4.49), then, from Lemma 4.4.1, it follows

$$\rho_3\parallel u^6\parallel_{L^2}^2+\parallel u^7\parallel_{L_g^2}^2\leq C\parallel U\parallel_{\mathcal{H}}\parallel F\parallel_{\mathcal{H}}. \tag{4.4.66}$$

From Lemma 4.4.2, for any $\varepsilon_2>0$, there exists a constant $C_1>0$ such that

$$\rho_2\parallel u^4\parallel_{L^2}^2\leq C_1\parallel F\parallel_{\mathcal{H}}\parallel U\parallel_{\mathcal{H}}+\frac{\bar{b}}{2}\parallel u_x^3\parallel_{L^2}^2+\frac{\varepsilon_2}{2}k\parallel u_x^1+u^3\parallel_{L^2}^2. \tag{4.4.67}$$

Also, from Lemma 4.4.3, we obtain

$$\bar{b} \parallel u_x^3 \parallel_{L^2}^2 \leq C_{\varepsilon_1} \parallel F \parallel_{\mathcal{H}} \parallel U \parallel_{\mathcal{H}} + \varepsilon_1 \rho_1 \parallel u^2 \parallel_{L^2}^2 + \varepsilon_2 k \parallel u_x^1 + u^3 \parallel_{L^2}^2 . \quad (4.4.68)$$

On the other hand, from Lemma 4.4.5, we have

$$k \parallel u_x^1 + u^3 \parallel_{L^2}^2 \leq C_3 \parallel F \parallel_{\mathcal{H}} \parallel U \parallel_{\mathcal{H}} + (\varepsilon_1 + \varepsilon_2)\rho_1 \parallel u^2 \parallel_{L^2}^2 . \quad (4.4.69)$$

By Lemma 4.4.5, we can derive

$$2(\varepsilon_1 + \varepsilon_2)\rho_1 \parallel u^2 \parallel_{L^2}^2 \leq 2(\varepsilon_1 + \varepsilon_2) \parallel F \parallel_{\mathcal{H}} \parallel U \parallel_{\mathcal{H}} + 8(\varepsilon_1 + \varepsilon_2) k \parallel u_x^1 + u^3 \parallel_{L^2}^2 . \quad (4.4.70)$$

Adding (4.4.66) and (4.4.67), we get

$$(1 - 8\varepsilon_1 + \varepsilon_2) k \parallel u_x^1 + u^3 \parallel_{L^2}^2 + (\varepsilon_1 + \varepsilon_2)\rho_1 \parallel u^2 \parallel_{L^2}^2 \leq C \parallel F \parallel_{\mathcal{H}} \parallel U \parallel_{\mathcal{H}} . \quad (4.4.71)$$

Finally, from Lemma 4.4.6, we get

$$\frac{\varepsilon_3 \beta}{4} \int_0^1 |u_x^5|^2 dx \leq C_{\varepsilon_3} \parallel U \parallel_{\mathcal{H}} \parallel F \parallel_{\mathcal{H}} + \frac{2\delta^2 \varepsilon_3}{\beta} \parallel u_x^4 \parallel^2 dx. \quad (4.4.72)$$

From (4.4.66), (4.4.67), (4.4.68), (4.4.71) and (4.4.72), we obtain that for $\varepsilon_1, \varepsilon_2, \varepsilon_3 > 0$ sufficiently small, there exists a constant $C > 0$, independent of λ, F, U, such that

$$\parallel U \parallel_{\mathcal{H}}^2 \leq C \parallel F \parallel_{\mathcal{H}}^2,$$

this completes the proof. □

4.5 Bibliographic Comments

For the results on global existence and asymptotic behavior of solutions to the system (4.3.1), we refer the reader to the bibliographic comments in Sect. 3.5.

Recently, Fernández Sare and Racke [81] considered hyperbolic Timoshenko-type vibrating systems that are coupled to a heat equation modeling an expectedly dissipative effect through heat conduction

$$\begin{cases} \rho_1 \varphi_{tt} - k(\varphi_x + \psi)_x = 0, \\ \rho_2 \psi_{tt} - b\psi_{xx} + \int_0^{+\infty} g(s)\psi_{xx}(t-s)ds + k(\varphi_x + \psi) + \delta\theta_x = 0, \\ \rho_3 \theta_t - q_x + \delta\psi_{xt} = 0, \\ \tau q_t + \beta q + \theta_x = 0 \end{cases} \quad (4.5.1)$$

with an exponential decaying kernel g, and further proved that system (4.5.1) is exponentially stable for the classical model, i.e., $\tau = 0$, but the exponential stability fails for the Cattaneo model, i.e., $\tau \neq 0$. Messaoudi et al. [174] solved the lack of exponential stability for the Cattaneo model by adding a linear damping of the form $\alpha\varphi_t$ to the left-hand side of the first equation in (4.5.1) under appropriate conditions of $\sigma, \rho_i, b, k, \gamma$ $(i = 1, 2, 3)$, Muñoz Rivera and Racke [191] considered the following system

$$\begin{cases} \rho_1\varphi_{tt} - \sigma(\varphi_x, \psi) = 0, & (x,t) \in (0,L) \times (0,+\infty), \quad (4.5.2)\\ \rho_2\psi_{tt} - b\psi_{xx} + k(\varphi_x + \psi) + \gamma\theta_x = 0, & (x,t) \in (0,L) \times (0,+\infty), \quad (4.5.3)\\ \rho_3\theta_t - k\theta_{xx} + \gamma\psi_{tx} = 0, & (x,t) \in (0,L) \times (0,+\infty), \quad (4.5.4) \end{cases}$$

and proved several exponential decay results for the linearized system and a nonexponential stability result for the case of different wave speeds. Messaoudi et al. [174] studied

$$\begin{cases} \rho_1\varphi_{tt} - \sigma(\varphi_x, \psi) + \mu\varphi_t = 0, & (x,t) \in (0,L) \times (0,+\infty), \quad (4.5.5)\\ \rho_2\psi_{tt} - b\psi_{xx} + k(\varphi_x + \psi) + \beta\theta_x = 0, & (x,t) \in (0,L) \times (0,+\infty), \quad (4.5.6)\\ \rho_3\theta_t + \gamma q_x + \delta\psi_{tx} = 0, & (x,t) \in (0,L) \times (0,+\infty), \quad (4.5.7)\\ \tau_0 q_t + q + k\theta_x = 0, \quad (x,t) \in (0,L) \times (0,+\infty), & \quad (4.5.8) \end{cases}$$

where $\varphi = \varphi(x,t)$ is the displacement vector, $\psi = \psi(x,t)$ is the rotation angle of the filament, $\theta(x,t)$ is the temperature difference, $q = q(x,t)$ is the heat flux vector, $\rho_1, \rho_2, \rho_3, b, k, \gamma, \delta, \mu, \tau_0$ are positive constants. The nonlinear function σ is assumed to be sufficiently smooth and satisfy

$$\sigma_{\varphi_x}(0,0) = \sigma_\psi(0,0) = k \tag{4.5.9}$$

and

$$\sigma_{\varphi_x\varphi_x}(0,0) = \sigma_{\varphi_x\psi}(0,0) = \sigma_{\psi\psi}(0,0) = 0. \tag{4.5.10}$$

Grobbelaar-Van Dalsen [94] studied the strong stabilization of models for the Reissner–Mindlin plate equations with second sound, that is, models that include thermal effects described according to Cattaneo's law of heat conduction instead of Fourier's law in classical thermoelasticity. Two models were considered which are distinct with respect to the property of compactness or non-compactness of the resolvent of the generator of the underlying semigroup. In accordance with the compactness or non-compactness of the resolvent operator, a different criterion for strong stability is implemented to achieve the strong stabilization of each model. Later on, Fastovska [72] established the relation between the thermoelastic Cattaneo model and the thermoelastic Gurtin–Pipkin model. The existence of the compact global attractor of the Cattaneo–Mindlin plate model was proved and its properties were studied.

Chapter 5
Global Existence of Solutions for the Thermoelastic Bresse System

5.1 Introduction

In this chapter, we shall show the global existence of solutions to the following (non)linear (non)homo-geneous thermoelastic Bresse systems by semigroup approaches

$$\begin{cases} \rho h w_{1tt} = (Eh(w_1' - kw_2) - \alpha\theta_1)' - kGh(\phi_1 + w_2' + kw_1) + f_1, & (5.1.1)\\ \rho h w_{2tt} = Gh(\phi_1 + w_2' + kw_1)' + kEh(w_1' - kw_2) - k\alpha\theta_1 + f_2, & (5.1.2)\\ \rho I \phi_{1tt} = EI\phi_1'' - Gh(\phi_1 + w_2' + kw_1) - \alpha\theta_1' + f_3, & (5.1.3)\\ \rho c\theta_{1t} = \theta_1'' - \alpha T_0(w_{1t}' - kw_{2t}) + f_4, & (5.1.4)\\ \rho c\theta_{2t} = \theta_2'' - \alpha T_0\phi_{1t}' + f_5, & (5.1.5) \end{cases}$$

with the boundary conditions

$$w'_1(t,x) = w_2(t,x) = \phi'_1(t,x) = \theta_1(t,x) = \theta_2(t,x) = 0, \ for\ x = 0, l, \quad (5.1.6)$$

or

$$w_1(t,x) = w_2(t,x) = \phi_1(t,x) = \theta_1(t,x) = \theta_2(t,x) = 0, \ for\ x = 0, l, \quad (5.1.7)$$

and initial conditions

$$w_1(0,x) = u_0(x),\ w_{1t}(0,x) = v_0(x),\ \phi_1(0,x) = \phi_0(x),\ \phi_{1t}(0,x) = \psi_0(x), \quad (5.1.8)$$

$$w_{2t}(0,x) = y_0(x),\ w_2(0,x) = w_0(x),\ \theta_1(0,x) = \theta_0(x),\ \theta_2(0,x) = \xi_0(x). \quad (5.1.9)$$

Y. Qin and Z. Ma, *Global Well-posedness and Asymptotic Behavior of the Solutions to Non-classical Thermo(visco)elastic Models*,
DOI 10.1007/978-981-10-1714-8_5

The total energy for this problem is given by

$$\mathcal{E}(t) = \frac{1}{2}\int_0^l \Big\{[Eh(w_1' - kw_2)^2 + Gh(\phi_1 + w_3' + kw_1)^2 + EI(\phi_1')^2] + [\rho h((w_{1t})^2 + (w_{2t})^2) + \rho I(\phi_{1t})^2] + \frac{\rho c}{T_0}\left(\theta_1^2 + \theta_2^2\right)\Big\}dx, \tag{5.1.10}$$

where the signs $'$ and $(\cdot)_t$, $(\cdot)_{tt}$ denote the partial derivatives with respect to spatial location $x \in [0, l]$ and time $t \geq 0$ respectively, w_1, w_2 and ϕ_1 are the longitudinal vertical and shear angle displacements; θ_1, θ_2 are the temperature deviations from the reference temperature T_0 along the longitudinal and vertical directions; $E, G, \rho, I, m, h, k, c$ are positive constants for the elastic and thermal material properties. We adopt the results in this chapter from [228].

From this seemingly complicated system (three wave equations coupled with two heat equations), very interesting special cases can be obtained. In particular, the isothermal system is exactly the system obtained by Bresse [21] in 1859. In their study on networks of flexible beams, Lagnese et al. [9] derived a general model for 3-d nonlinear thermoelastic beams. A special case of this model is a linear planar, shearable thermoelastic beam whose motion is governed by the following system of partial differential equations

$$\begin{cases} \rho h w_{1tt} &= (Eh(w_1' - kw_3) - \alpha\theta_1)' - kGh(\phi_2 + w_3' + kw_1), \\ \rho h w_{3tt} &= Gh(\phi_2 + w_3' + kw_1)' + kEh(w_1' - kw_3) - k\alpha\theta_1, \\ \rho I\phi_{2tt} &= EI\phi_2'' - Gh(\phi_2 + w_3' + kw_1) - \alpha\theta_3', \\ \rho c\theta_{1t} &= \theta_1'' - \alpha T_0(w_{1t}' - kw_{3t}), \\ \rho c\theta_{3t} &= \theta_3'' - \alpha T_0\phi_{2t}', \end{cases}$$

where w_1, w_2 and ϕ_2 are the longitudinal vertical and shear angle displacements; θ_1, θ_3 are the temperature deviations from the reference temperature T_0 along the longitudinal and vertical directions; $E, G, \rho, I, m, h, k, c$ are positive constants for the elastic and thermal material properties.

The Bresse system, Eqs. (5.1.1)–(5.1.3) with θ_1, θ_2 removed, is more general than the well-known Timoshenko system where the longitudinal displacement w_1 is not considered. If both θ_1 and w_1 are neglected, the Bresse thermoelastic system reduces to the following Timoshenko thermoelastic system

$$\begin{cases} \rho h w_{2tt} &= Gh(\phi_1 + w_2' + kw_1)' + kEh(w_1' - kw_2) - k\alpha\theta_1, \\ \rho I\phi_{1tt} &= EI\phi_1'' - Gh(\phi_1 + w_2' + kw_1) - \alpha\theta_1', \\ \rho c\theta_{2t} &= \theta_2'' - \alpha T_0\phi_{1t}'. \end{cases}$$

5.2 Main Results

In order to simplify the notation, we put

$$u = w_1,\ w = w_3,\ \phi = \phi_2,\ v = w_{1t},\ y = w_{2t},\ \psi = \phi_{1t},\ \theta = \theta_1,\ \xi = \theta_2$$

and set the state variable vector

$$\Phi = (u, w, \phi, v, y, \psi, \theta, \xi). \tag{5.2.1}$$

In order to choose the proper state space for the system, we shall find the static solution first. Thus we consider the static system associated with problem (4.1.1)–(4.1.5)

$$\begin{cases} Eh(u_1' - kw)' + \alpha\theta' - kGh(\phi + w' + ku)' = 0, & (5.2.2)\\ Gh(\phi + w' + ku)' + kEh(u' - kw) - k\alpha\theta = 0, & (5.2.3)\\ EI\phi'' - Gh(\phi + w' + ku) - \alpha\xi' = 0, & (5.2.4)\\ \theta'' = 0, & (5.2.5)\\ \xi'' = 0. & (5.2.6) \end{cases}$$

Since $\theta,\ \xi$ both vanish at $x = 0$ or l, it follows from the Eqs. (5.2.5)–(5.2.6) that $\theta = \xi = 0$. Then we multiply (5.2.2)–(5.2.6) by $u,\ w,\ \phi$ and integrate from 0 to l, respectively. This yields

$$\int_0^l \left[Eh(u' - kw)^2 + Gh(\phi + w' + ku)^2 + EI(\phi')^2\right](x)dx = 0. \tag{5.2.7}$$

Similarly to the argument in Liu and Rao [151], we assume that $g(t) = \int_0^l [\phi(x, t) + ku(x, t)]dx$ and choose the state spaces

$$\begin{cases} \mathcal{H}_1 = H_*^1 \times H_0^1 \times H_*^1 \times L_*^2 \times L^2 \times L_*^2 \times (L^2)^2, & (5.2.8)\\ \mathcal{H}_2 = (H_0^1)^3 \times (L^2)^5, & (5.2.9) \end{cases}$$

where

$$H_*^1 = \left\{g \in H^1(0, l) | \int_0^l g(x)dx = 0\right\},\ L_*^2 = \left\{g \in L^2(0, l) | \int_0^l g(x)dx = 0\right\}. \tag{5.2.10}$$

Both state spaces are equipped with proper inner products on $\mathcal{H}_i$

$$\begin{aligned}(z_1, z_2)_{\mathcal{H}_i} &= Eh(u_1' - kw_1, u_2' - kw_2) + Gh(\phi_1 + w_1' + ku_1, \phi_2 + w_2' + ku_2)\\ &+ EI(\phi_1', \phi_2') + \rho h(v_1, v_2) + \rho h(y_1, y_2)\\ &+ \rho I(\psi_1, \psi_2) + \frac{\rho c}{T_0}\Big[(\theta_1, \theta_2) + (\xi_1, \xi_2)\Big], \quad i = 1, 2,\end{aligned}$$

for all $z_1 = (u_1, w_1, \phi_1, v_1, y_2, \psi_1, \theta_1, \xi_1)$, $z_2 = (u_2, w_2, \phi_2, v_2, y_2, \psi_2, \theta_2, \xi_2) \in \mathcal{H}_i$ $(i = 1, 2)$, which induce the energy norms

$$\begin{aligned}\|z\|^2_{\mathcal{H}_i} &= Eh\|u' - kw\|^2 + Gh\|\phi + w' + ku\|^2 + EI\|\phi'\|^2 + \rho h\|v\|^2 + \rho h\|y\|^2\\ &+ \rho I\|\psi\|^2 + \frac{\rho c}{T_0}(\|\theta\|^2 + \|\xi\|^2), \quad i = 1, 2,\end{aligned} \tag{5.2.11}$$

for $z = (u, w, \phi, v, y, \psi, \theta, \xi) \in \mathcal{H}_i$ $(i = 1, 2)$, and $(\cdot, \cdot)$ denotes the inner product on $L^2(0, l)$, $\|\cdot\|$ stands for the $L^2(0, l)$ norm.

Define linear operators $A_j : \mathcal{H}_j \to \mathcal{H}_j$ $(j = 1, 2)$ by

$$A_j\Phi = \left(v, y, \psi, B, C, D, \frac{1}{\rho c}\theta'' - \frac{\alpha T_0}{\rho c}(v' - ky), \frac{1}{\rho c}\xi'' - \frac{\alpha T_0}{\rho c}\psi'\right), \tag{5.2.12}$$

where

$$\begin{cases} B = B(u, w, \phi, \theta) = \dfrac{E}{\rho}(u' - kw)' - \dfrac{\alpha}{\rho h}\theta' - \dfrac{kG}{\rho}(\phi + w' + ku), & (5.2.13)\\ C = C(u, w, \phi, \theta) = \dfrac{G}{\rho}(\phi + w' + ku)' + \dfrac{kE}{\rho}(u' - kw) - \dfrac{k\alpha}{\rho h}\theta, & (5.2.14)\\ D = D(u, w, \phi, \xi) = \dfrac{E}{\rho}\phi'' - \dfrac{Gh}{\rho I}(\phi + w' + ku) - \dfrac{\alpha}{\rho I}\xi', & (5.2.15)\end{cases}$$

and

$$\begin{cases} \mathcal{D}(A_1) = \{a \in \mathcal{H}_1 | w, \theta, \xi \in H_0^1 \cap H^2, u', \phi', y \in H_0^1, v, \psi \in H_*^1\}, & (5.2.16)\\ \mathcal{D}(A_2) = \{a \in \mathcal{H}_2 | u, w, \phi, \theta, \xi \in H_0^1 \cap H^2, v, y, \psi \in H_0^1\}. & (5.2.17)\end{cases}$$

Set

$$\Phi = (u, w, \phi, v, y, \psi, \theta, \xi), K = (0, 0, 0, f_1, f_2, f_3, f_4, f_5), \tag{5.2.18}$$

thus the nonhomogeneous thermoelastic Bresse beam system can be transformed to a first order nonhomogeneous evolutionary equation on the Hilbert spaces $\mathcal{H}_j$:

$$\Phi_t = A_j\Phi + K, \Phi(0) = \Phi_0 = (u_0, w_0, \phi_0, v_0, y_0, \psi_0, \theta_0, \xi_0) \tag{5.2.19}$$

with $j = 1, 2$ corresponding to the boundary conditions (5.1.6)–(5.1.7), respectively.

Our hypothesis on f_i $(i = 1, 2, 3, 4, 5)$ are as follows for different theorems below:

$$\begin{cases} (H1)\, f_i &= 0; \\ (H2)\, f_i &= f_i(t, x) \in C^1([0, +\infty), L^2(0, l)); \\ (H3)\, f_i &= f_i(t, x) \in C([0, +\infty), H_0^1(0, l)); \\ (H4)\, f_i &= f_i(\Phi(t, x)) \in C([0, +\infty), L^2(0, l)),\ f_{it} \in L^1((0, T), L^2(0, l))\ \textit{for any}\ T > 0, \end{cases}$$

where $\Phi(t, s)$ is a function defined on $H_j, j = 1, 2$.

Theorem 5.2.1 (Homogeneous Case) *Suppose that the assumption* $(H1)$ *holds, then for any* $\Phi(0) = (u_0, w_0, \phi_0, v_0, y_0, \psi_0, \theta_0, \xi_0) \in D(A_1)$, *there exists a unique global classical solution* $\Phi = (u, w, \phi, v, y, \psi, \theta, \xi)$ *to system (5.1.1)–(5.1.9) satisfying*

$$\Phi = (u, w, \phi, v, y, \psi, \theta, \xi) \in C^1([0, +\infty), \mathcal{H}_1) \cap C([0, +\infty), D(A_1)), \quad (5.2.20)$$

i.e.,

$$\begin{cases} u & \in C^1([0, +\infty), H_*^1(0, l)) \cap C([0, +\infty), H^2(0, l) \cap H_0^1(0, l)), \\ w & \in C^1([0, +\infty), H_0^1(0, l)) \cap C([0, +\infty), H^2(0, l) \cap H_0^1(0, l)), \\ \phi & \in C^1([0, +\infty), H_*^1(0, l)) \cap C([0, +\infty), H^2(0, l) \cap H_0^1(0, l)), \\ v & \in C^1([0, +\infty), L_*^2(0, l)) \cap C([0, +\infty), H_*^1(0, l)), \\ y & \in C^1([0, +\infty), L^2(0, l)) \cap C([0, +\infty), H_0^1(0, l)), \\ \psi & \in C^1([0, +\infty), L_*^2(0, l)) \cap C([0, +\infty), H_*^1(0, l)), \\ \theta & \in C^1([0, +\infty), L^2(0, l)) \cap C([0, +\infty), H^2(0, l) \cap H_0^1(0, l)), \\ \xi & \in C^1([0, +\infty), L^2(0, l)) \cap C([0, +\infty), H^2(0, l) \cap H_0^1(0, l)). \end{cases}$$

Theorem 5.2.2 (Nonhomogeneous Case) *Suppose that the hypothesis* $(H2)$ *holds, then for any* $\Phi(0) = (u_0, w_0, \phi_0, v_0, y_0, \psi_0, \theta_0, \xi_0) \in D(A_1)$, *there exists a unique global classical solution* $\Phi = (u, w, \phi, v, y, \psi, \theta, \xi)$ *to system (5.1.1)–(5.1.9) satisfying*

$$\Phi = (u, w, \phi, v, y, \psi, \theta, \xi) \in C^1([0, +\infty), \mathcal{H}_1) \cap C([0, +\infty), D(A_1)). \quad (5.2.21)$$

Theorem 5.2.3 (Semilinear Case) *Suppose that the condition* $(H3)$ *holds,* $f_i = f_i(\Phi)$, $\Phi = (u, w, \phi, v, y, \psi, \theta, \xi)$. $K = (0, 0, 0, f_1, f_2, f_3, f_4, f_5)$ *in (4.2.18) satisfies the global Lipschitz condition on* $\mathcal{H}_1$, *i.e., there exists a positive constant L such that for all* $\Phi_1, \Phi_2 \in \mathcal{H}_1$,

$$\|K(\Phi_1) - K(\Phi_2)\|_{\mathcal{H}_1} \le L\|\Phi_1 - \Phi_2\|_{\mathcal{H}_1}, \quad (5.2.22)$$

then for any $\Phi(0) = (u_0, w_0, \phi_0, v_0, y_0, \psi_0, \theta_0, \xi_0) \in \mathcal{H}_1$, *there exists a unique global mild solution* $\Phi = (u, w, \phi, v, y, \psi, \theta, \xi)$ *to system (5.1.1)–(5.1.9) such that*

$$\Phi \in C([0, +\infty), \mathcal{H}_1), \tag{5.2.23}$$

i.e.,

$$\begin{cases} u & \in C^1([0, +\infty), H_*^1(0, l)), \ w \in C^1([0, +\infty), H_0^1(0, l)), \\ \phi & \in C^1([0, +\infty), H_*^1(0, l)), \ v \in C^1([0, +\infty), L_*^2(0, l)), \\ y & \in C^1([0, +\infty), L^2(0, l)), \ \psi \in C^1([0, +\infty), L_*^2(0, l)), \\ \theta & \in C^1([0, +\infty), L^2(0, l)), \ \xi \in C^1([0, +\infty), L^2(0, l)). \end{cases}$$

Theorem 5.2.4 (Nonlinear Case) *Suppose the assumption* $(H4)$ *holds,* $f_i = f_i(\Phi)$, $K = (0, 0, 0, f_1, f_2, f_3, f_4, f_5)$ *in (5.2.18) is a nonlinear operator from* $D(A_1)$ *to* $D(A_1)$, *and satisfies the global Lipschitz condition on* $D(A_1)$, *i.e., there exists a positive constant* L *such that for all* $\Phi_1, \ \Phi_2 \in D(A_1)$,

$$\|K(\Phi_1) - K(\Phi_2)\|_{D(A_1)} \leq L\|\Phi_1 - \Phi_2\|_{D(A_1)}. \tag{5.2.24}$$

Then for any $\Phi(0) = (u_0, w_0, \phi_0, v_0, y_0, \psi_0, \theta_0, \xi_0) \in D(A_1)$, *there exists a unique global classical solution* Φ *to system (5.1.1)–(5.1.9) such that*

$$\Phi = (u, w, \phi, v, y, \psi, \theta, \xi) \in C^1([0, +\infty), \mathcal{H}_1) \cap C([0, +\infty), D(A_1)). \tag{5.2.25}$$

Similarly to the case of $A_j = A_1$, we can show the following remark.

Remark 5.2.1 Theorems 5.2.1–5.2.4 also hold for the case $j = 2$.

5.3 Some Lemmas

In order to complete the proofs of Theorems 5.2.1–5.2.4, we shall use Lemmas 1.3.14–1.3.18 and the following lemmas.

To prove that the operator A_1 defined by (5.2.12) is dissipative, we need the following lemma.

Lemma 5.3.1 *The operator* A_1 *defined by (5.2.12) is dissipative. Furthermore,* $0 \in \rho(A_1)$, *where* $\rho(A_1)$ *is the resolvent of the operator* A_1.

Proof By a straightforward calculation, we derive

$$\begin{aligned} &\langle A_1(u, w, \phi, v, y, \psi, \theta, \xi), (u, w, \phi, v, y, \psi, \theta, \xi)\rangle_{\mathcal{H}_1} \\ &= (v, u)_{H_*^1} + (y, w)_{H_0^1} + (\psi, \phi)_{H_*^1} + \left(\frac{E}{\rho}(u' - kw)' - \frac{\alpha}{\rho h}\theta' - \frac{kG}{\rho}(\phi + w' + ku), v\right)_{L_*^2} \\ &+ \left(\frac{G}{\rho}(\phi + w' + ku)' + \frac{kE}{\rho}(u' - kw) - \frac{k\alpha}{\rho h}\theta, y\right)_{L_*^2} \\ &+ \left(\frac{E}{\rho}\phi'' - \frac{Gh}{\rho I}(\phi + w' + ku) - \frac{\alpha}{\rho I}\xi', \psi\right)_{L^2} \end{aligned}$$

$$+\left(\frac{1}{\rho c}\theta'' - \frac{\alpha T_0}{\rho c}\left(v' - ky\right), \theta\right)_{L^2} + \left(\frac{1}{\rho c}\xi'' - \frac{\alpha T_0}{\rho c}\psi', \xi\right)_{L^2}$$
$$= -\frac{1}{T_0}\left(\|\theta_t\|^2 + \|\xi_t\|^2\right) \le 0. \tag{5.3.1}$$

Thus, A_1 is dissipative.

Next, we shall prove that $0 \in \rho(A_1)$. For any $G = (g_1, g_2, g_3, g_4, g_5, g_6, g_7, g_8) \in \mathcal{H}_1$, we consider the abstract equation

$$A_1\Phi = G, \tag{5.3.2}$$

i.e.,

$$\begin{cases} v = g_1, & (5.3.3)\\ y = g_2, & (5.3.4)\\ \psi = g_3, & (5.3.5)\\ \frac{E}{\rho}\left(u' - kw\right)' - \frac{\alpha}{\rho h}\theta' - \frac{kG}{\rho}\left(\phi + w' + ku\right) = g_4, & (5.3.6)\\ \frac{G}{\rho}\left(\phi + w' + ku\right)' + \frac{kE}{\rho}\left(u' - kw\right) - \frac{k\alpha}{\rho h}\theta = g_5, & (5.3.7)\\ \frac{E}{\rho}\phi'' - \frac{Gh}{\rho I}\left(\phi + w' + ku\right) - \frac{\alpha}{\rho I}\xi' = g_6, & (5.3.8)\\ \frac{1}{\rho c}\theta'' - \frac{\alpha T_0}{\rho c}\left(v' - ky\right) = g_7, & (5.3.9)\\ \frac{1}{\rho c}\xi'' - \frac{\alpha T_0}{\rho c}\psi' = g_8. & (5.3.10) \end{cases}$$

Clearly, (5.3.5) has a unique solution $\Phi \in \mathcal{D}(A_1)$. In fact, using the standard elliptic equation theory, we have

$$\theta \in H_0^1 \cap H^2, \quad \xi \in H_0^1 \cap H^2. \tag{5.3.11}$$

Define a bilinear operator b as

$$b((u, w, \phi), (\widetilde{u}, \widetilde{w}, \widetilde{\phi}))$$
$$= Eh(u' - kw, \widetilde{u}' - k\widetilde{w}') + Gh(\phi + w' + ku, \widetilde{\phi} + \widetilde{w}' + k\widetilde{u}) + EI(\phi', \widetilde{\phi}'), \tag{5.3.12}$$

and using the Lax–Milgram theorem for the Eqs. (5.3.6)–(5.3.13) with the corresponding variables (u, w, ϕ), we can easily deduce the conclusion, i.e., $0 \in \rho(A_1)$. □

Lemma 5.3.2 *The operator A_1 defined by (5.2.12) is closed.*

Proof Choose $(u_n, w_n, \phi_n, v_n, y_n, \psi_n, \theta_n, \xi_n) \in D(A_1)$ such that

$$(u_n, w_n, \phi_n, v_n, y_n, \psi_n, \theta_n, \xi_n) \to (u, w, \phi, v, y, \psi, \theta, \xi) \quad \text{in } \mathcal{H}_1$$

and

$$A_1(u_n, w_n, \phi_n, v_n, y_n, \psi_n, \theta_n, \xi_n) \to (a_1, a_2, a_3, a_4, a_5, a_6, a_7, a_8) \quad \text{in } \mathcal{H}_1. \tag{5.3.13}$$

That is, we have

$$\begin{cases} u_n \to u \quad \text{in } H^1_*(0,l), & (5.3.14)\\ w_n \to w \quad \text{in } H^1_0(0,l), & (5.3.15)\\ \phi_n \to \phi \quad \text{in } H^1_*(0,l), & (5.3.16)\\ v_n \to v \quad \text{in } L^2_*(0,l), & (5.3.17)\\ y_n \to y \quad \text{in } L^2_*(0,l), & (5.3.18)\\ \psi_n \to \psi \quad \text{in } L^2(0,l), & (5.3.19)\\ \theta_n \to \theta \quad \text{in } L^2(0,l), & (5.3.20)\\ \xi_n \to \xi \quad \text{in } L^2(0,l), & (5.3.21) \end{cases}$$

and

$$\begin{cases} v_n \to a_1 \quad \text{in } H^1_*(0,l), & (5.3.22)\\ y_n \to a_2 \quad \text{in } H^1_0(0,l), & (5.3.23)\\ \psi_n \to a_3 \quad \text{in } H^1_*(0,l), & (5.3.24)\\ \frac{E}{\rho}\left(u_n' - kw_n\right)' - \frac{\alpha}{\rho h}\theta_n' - \frac{kG}{\rho}\left(\phi_n + w_n' + ku_n\right) \to a_4 \quad \text{in } L^2_*(0,l), & (5.3.25)\\ \frac{G}{\rho}\left(\phi_n + w_n' + ku_n\right)' + \frac{kE}{\rho}\left(u_n' - kw_n\right) - \frac{k\alpha}{\rho h}\theta_n \to a_5 \quad \text{in } L^2_*(0,l), & (5.3.26)\\ \frac{E}{\rho}\phi_n'' - \frac{Gh}{\rho I}\left(\phi_n + w_n' + ku_n\right) - \frac{\alpha}{\rho I}\xi_n' \to a_6 \quad \text{in } L^2(0,l), & (5.3.27)\\ \frac{1}{\rho c}\theta_n'' - \frac{\alpha T_0}{\rho c}\left(v_n' - ky_n\right) \to a_7 \quad \text{in } L^2(0,l), & (5.3.28)\\ \frac{1}{\rho c}\xi_n'' - \frac{\alpha T_0}{\rho c}\psi_n' \to a_8 \quad \text{in } L^2(0,l). & (5.3.29) \end{cases}$$

By (5.3.21) and (5.3.26), we deduce

$$v_n \to v \;\; in \;\; H^1_*(0,l) \tag{5.3.30}$$

and

$$v = a_1 \quad \in H^1_*(0,l). \tag{5.3.31}$$

By (5.3.22) and (5.3.27), we derive

$$y_n \to y \ \ in\, H_0^1(0, l) \cap L_*^2(0, l) \tag{5.3.32}$$

and

$$y = a_2 \ \ \in\ H_0^1(0, l) \cap L_*^2(0, l). \tag{5.3.33}$$

By (5.3.23) and (5.3.28), we obtain

$$\psi_n \to \psi \ \ in\, H_*^1(0, l) \tag{5.3.34}$$

and

$$\psi = a_3 \ \ \in H_*^1(0, l). \tag{5.3.35}$$

By (5.3.25), (5.3.33), (5.3.37) and (5.3.38), we have

$$\xi_n \to \xi \ \ in\, H^2(0, l). \tag{5.3.36}$$

By (5.3.24), (5.3.32) and (5.3.34)–(5.3.37), we deduce

$$\theta_n \to \theta \ \ in\, H^2(0, l). \tag{5.3.37}$$

From (5.3.18)–(5.3.20), (5.3.31) and (5.3.39)–(5.3.40), we derive

$$\phi_n \to \phi \ \ in\, H^2(0, l) \cap H_*^1(0, l). \tag{5.3.38}$$

From (5.3.18)–(5.3.19), (5.3.29)–(5.3.30) and (5.3.34)–(5.3.41), we get

$$u_n \to u \ \ in\ \ H^2(0, l) \cap H_*^1(0, l) \tag{5.3.39}$$

and

$$w_n \to w \ \ in\ \ H^2(0, l) \cap H_0^1(0, l). \tag{5.3.40}$$

Thus from (5.3.34)–(5.3.40), we derive

$$A_1(u, w, \phi, v, y, \psi, \theta, \xi) = (a_1, a_2, a_3, a_4, a_5, a_6, a_7, a_8), \ \ (u, w, \phi, v, y, \psi, \theta, \xi) \in D(A_1),$$

that is, A_1 is closed. □

Lemma 5.3.3 *Let A_1 be a densely defined linear operator on a Hilbert space $\mathcal{H}_1$. Then A_1 generates a C_0-semigroup of contractions on $\mathcal{H}_1$ if and only if A_1 is dissipative and $R(\lambda I - A_1) = \mathcal{H}_1$.*

Proof By the resolvent identity, for small $\lambda > 0$, we have $R(\lambda I - A_1) = \mathcal{H}_1$. From Lemmas 5.3.1 and 5.3.2, 1.3.13 and Lumer–Phillip theorem, the sufficiency is obvious. For more details, we can refer to Liu and Zheng [155] or Pazy [201]. □

5.4 Proofs of Theorems 5.2.1–5.2.4

Proofs of Theorems 5.2.1 By (5.2.8), it is clear that $\mathcal{H}_1$ is a Hilbert space. By Lemmas 1.3.13 and 5.3.3, we can deduce that the operator A_1 is the infinitesimal generator of a contractive C_0-semigroup on Hilbert space $\mathcal{H}_1$. Applying Lemma 1.3.14, we can obtain our result. □

Proofs of Theorems 5.2.2 We have known that the operator A_1 is the infinitesimal generator of a contractive C_0-semigroup on Hilbert space $\mathcal{H}_1$. Applying Lemma 5.3.3, we can obtain $R(\lambda I - A_1) = \mathcal{H}_1$. Choosing the operator $B = -A_1$, we can obtain $D(B) = D(A_1)$ and $D(B)$ is dense in $\mathcal{H}_1$. Noting that $K = (0, 0, 0, f_1, f_2, f_3, f_4, f_5) \in [C^1([0, +\infty), L^2(0, l))]^8$ and applying Definition 1.3.6, we can conclude that the operator B is a maximal accretive operator. Then we can complete the proof of Theorem 5.2.2 by virtue of Lemmas 1.3.15–1.3.16. □

Proofs of Theorems 5.2.3 By virtue of the proof of Theorem 5.2.2, we know that $B = -A_1$ is a maximal accretive operator of a C_0 semigroup $S(t)$. On the other hand, $K = (0, 0, 0, f_1, f_2, f_3, f_4, f_5)$ satisfies the global Lipschitz condition on $\mathcal{H}_1$.
Let

$$P(\Phi) = S(t)\Phi_0 + \int_0^t S(t-\tau)K(\Phi(\tau))d\tau \tag{5.4.1}$$

and

$$W = \left\{ \Phi \in C([0, +\infty), \mathcal{H}_1) \mid \sup_{t\geq 0}(\| \Phi(t) \| e^{-kt}) < +\infty \right\}, \tag{5.4.2}$$

where k is a positive constant satisfying $k > L$. In W, we introduce the following norm:

$$\| \Phi \|_W = \sup_{t\geq 0}\Big(\| \Phi(t) \| e^{-kt}\Big). \tag{5.4.3}$$

Clearly, W is a Banach space.

Next, we shall use the contractive mapping theorem to prove our theorem. We shall show that the nonlinear operator P defined by (5.4.1) maps W into itself, and the mapping is contractive on W.
Indeed, for $\Phi \in W$, we have

$$
\begin{aligned}
\| P(\Phi) \| &\le \| S(t)\Phi_0 \| + \int_0^t \| S(t-\tau) \| \| K(\Phi) \| \, d\tau \\
&\le \| \Phi_0 \| + \int_0^t \| K(\Phi) \| \, d\tau \le \| \Phi_0 \| + \int_0^t (L \| \Phi(\tau) \| + \| K(0) \|) d\tau \\
&\le \| \Phi_0 \| + C_0 t + L \sup_{t\ge 0} \| \Phi(t) \| e^{-kt} \int_0^t e^{k\tau} d\tau \\
&\le \| \Phi_0 \| + C_0 t + \frac{L}{k} e^{kt} \| \Phi \|_W
\end{aligned}
\tag{5.4.4}
$$

where $C_0 = \| K(0) \|$.

Thus,

$$
\| P(\Phi) \|_W \le \sup_{t\ge 0} \left((\| \Phi_0 \| + C_0 t) e^{-kt} \right) + \frac{L}{k} \| \Phi \|_W < +\infty, \tag{5.4.5}
$$

which means $P(\Phi) \in W$.

For $\Phi_1, \ \Phi_2 \in W$, we have

$$
\begin{aligned}
\| P(\Phi_1) - P(\Phi_2) \|_W &= \sup_{t\ge 0} e^{-kt} \left\| \int_0^t S(t-\tau)(K(\Phi_1(\tau)) - K(\Phi_2(\tau))) d\tau \right\| \\
&\le \sup_{t\ge 0} e^{-kt} L \int_0^t \| \Phi_1 - \Phi_2 \| \, d\tau \le \sup_{t\ge 0} \left(e^{-kt} \cdot \frac{L}{k} \cdot (e^{kt} - 1) \right) \| \Phi_1 - \Phi_2 \|_W \\
&\le \frac{L}{k} \| \Phi_1 - \Phi_2 \|_W \, .
\end{aligned}
\tag{5.4.6}
$$

Therefore, by the contraction mapping theorem, the problem (5.2.19) has a unique solution in W. Let $\Phi_1, \ \Phi_2 \in C([0, +\infty), \mathcal{H}_1)$ be two solutions of the problem, set $\Phi = \Phi_1 - \Phi_2$, then we have

$$
\Phi(t) = \int_0^t S(t-\tau) \left(K(\Phi_1) - K(\Phi_2) \right) d\tau, \tag{5.4.7}
$$

$$
\| \Phi(t) \| \le L \int_0^t \| \Phi(\tau) \| \, d\tau. \tag{5.4.8}
$$

By the Bellman–Gronwall inequality, we conclude that $\Phi(t) = 0$, i.e., the uniqueness in $C([0, +\infty), \mathcal{H}_1)$ follows. Thus the proof is complete. □

Proofs of Theorems 5.2.4 Since B is a maximal accretive operator, $K = (0, 0, 0, f_1, f_2, f_3, f_4, f_5)$ satisfies the global Lipschitz condition on $D(A_1)$, let

$$
\mathcal{H}_1 = D(B), \ \ B_1 = B^2 : \ \ D(B_1) = D(B^2) \to \mathcal{H}_1, \tag{5.4.9}
$$

then $\mathcal{H}_1$ is a Banach space, and $B_1 = B^2$ is a densely defined operator from $D(B^2)$ into $\mathcal{H}_1$.
Next, we shall prove that B_1 is m-accretive in $\mathcal{H}_1 = D(B)$. For any $x,\ y \in D(B^2)$, since B is accretive in $\mathcal{H}_1$, we have

$$\begin{aligned}
&\| x - y + \lambda(Bx - By) \|_{D(B)} \\
&= \Big(\| x - y + \lambda(Bx - By) \|^2 + \| Bx - By + \lambda(B^2x - B^2y) \|^2 \Big)^{\frac{1}{2}} \\
&\geq \Big(\| x - y \|^2 + \lambda \| Bx - By \|^2 \Big)^{\frac{1}{2}} = \| x - y \|_{D(B)},
\end{aligned} \tag{5.4.10}$$

i.e., B_1 is accretive in $\mathcal{H}_1$. Furthermore, since B is m-accretive in $\mathcal{H}_1$, for any $y \in \mathcal{H}_1$, there exists a unique $x \in D(B)$ such that

$$x + Bx = y. \tag{5.4.11}$$

For any $y \in \mathcal{H}_1 = D(B)$, Eq. (5.4.11) admits a unique solution $x \in D(B)$, which implies

$$Bx = y - x \ \in \ D(B), \tag{5.4.12}$$

thus $x \in D(B^2)$, i.e., B_1 is m-accretive in $\mathcal{H}_1$.
Let $S_1(t)$ be the semigroup generated by B_1. If $\Phi_0 \in \ D(B^2) = D(B_1)$, then

$$\Phi(t) = S_1(t)\Phi_0 \in C([0, +\infty), D(B^2)) \cap C^1([0, +\infty), D(B))$$

is a unique classical solution of the problem. On the other hand, $\Phi(t) = S_1(t)\Phi_0$ is also a classical solution in

$$C([0, +\infty), D(B)) \cap C^1([0, +\infty), \mathcal{H}_1),$$

which implies that $S_1(t)$ is a restriction of $S(t)$ on $\mathcal{H}_1$.
Similarly to the proof of Theorem 5.2.3, there exists a unique solution $\Phi \in C([0, +\infty), \mathcal{H}_1)$. Since $S_1(t)$ is a restriction of $S(t)$ on $D(B)$, $K(\Phi)$ is an operator from $D(B)$ to $D(B)$, by Lemmas 1.3.16, 1.3.18, we conclude that Φ is a classical solution to the problem. Thus the proof is complete. □

5.5 Bibliographic Comments

When $f_i = 0$ $(i = 1, 2, 3, 4, 5)$, system (5.1.1)–(5.1.5) reduces to the following homogeneous thermoelastic Bresse system

$$
\begin{cases}
\rho h w_{1tt} = (Eh(w_1' - kw_3) - \alpha\theta_1)' - kGh(\phi_2 + w_3' + kw_1), & (5.5.1)\\
\rho h w_{3tt} = Gh(\phi_2 + w_3' + kw_1)' + kEh(w_1' - kw_3) - k\alpha\theta_1, & (5.5.2)\\
\rho I \phi_{2tt} = EI\phi_2'' - Gh(\phi_2 + w_3' + kw_1) - \alpha\theta_3', & (5.5.3)\\
\rho c\theta_{1t} = \theta_1'' - \alpha T_0(w_{1t}' - kw_{3t}), & (5.5.4)\\
\rho c\theta_{3t} = \theta_3'' - \alpha T_0\phi_{2t}'. & (5.5.5)
\end{cases}
$$

The well-known Timoshenko system is a special case of Bresse system when θ_1, θ_3 in the Eqs. (5.5.1)–(5.5.5) are removed while the longitudinal displacement w_1 is not considered:

$$
\begin{cases}
hw_{3tt} = Gh(\phi_2 + w_{3x})_x, & (5.5.6)\\
\rho I\phi_{2tt} = EI\phi_{2xx} - Gh(\phi_2 + w_{3x}) - \alpha\theta_{3tx}, & (5.5.7)\\
\rho c\theta_{3tt} = \theta_{3xxt} + \theta_{3xx} - \alpha T_0\phi_{2tx}, & (5.5.8)
\end{cases}
$$

which has been studied by Messaoudi and Said-Houari [175].

The researches on thermoelastic Bresse system have attracted many mathematicians (see, e.g., [118, 151, 157, 205, 244, 288]). Bresse [21] obtained the isothermal Bresse system in 1856. Recently, Liu and Rao [151] investigated the exponential decay and polynomial decay rates of the case $E = G$ and $E \neq G$ respectively for the homogeneous Bresse system when the wave speed of the vertical displacement coincides with the wave speed of longitudinal displacement or of the shear angle displacement. Ma [157] proved the exponential decay and global attractors for (5.5.1)–(5.5.5) by multiplier techniques. Fatori and Muñoz Rivera [75] considered the Bresse system with temperature and showed that, in general, the system is not exponentially stable but that there exists polynomial stability with rates that depend on the wave propagations and the regularity of the initial data. Moreover, they introduced a necessary condition to dissipative semigroup decay polynomially. Later on, Najdi and Wehbe [195] extended their results, by taking into consideration the important case when the thermal dissipation is locally distributed and to improve the polynomial energy decay rate, they then studied the energy decay rate of Bresse system with one locally thermal dissipation law, under the equal speed wave propagation condition, they established an exponential energy decay rate, on the contrary, they proved that the energy of the system decays, in general, at the rate $t^{(-1/2)}$. Moreover, using the frequency-domain characterization, the exponential decay has been obtained by Prüss [205] and Huang [118] in different cases. Recently, Dell'Oro [54] provided a comprehensive stability analysis of the thermoelastic Bresse system, in particular, assuming a temperature evolution of Gurtin–Pipkin type, the authors established a necessary and sufficient condition for exponential stability in terms of the structural parameters of the problem. As a byproduct, a complete characterization of the longtime behavior of Bresse-type systems with Fourier, Maxwell–Cattaneo and Coleman–Gurtin thermal laws was obtained, their main theorem also subsumed some recent achievements in the stability properties of thermoelastic Timoshenko systems with classical and nonclassical heat conduction.

Chapter 6
Stability for a Thermoelastic Bresse System

6.1 Introduction

In this chapter, we shall use the multiplier techniques to prove the exponential stability result only for equal wave speeds for the following thermoelastic Bress system

$$
\begin{cases}
\rho h w_{1tt} = (Eh(w_{1x} - kw_3) - \alpha\theta_1)_x - kGh(\phi_2 + w_{3x} + kw_1), & (6.1.1)\\
\rho h w_{3tt} = Gh(\phi_2 + w_{3x} + kw_1)_x + kEh(w_{1x} - kw_3) - k\alpha\theta_1, & (6.1.2)\\
\rho I \phi_{2tt} = EI\phi_{2xx} - Gh(\phi_2 + w_{3x} + kw_1) - \alpha\theta_{3xt}, & (6.1.3)\\
\rho c\theta_{1t} = \theta_{1xx} - \alpha T_0(w_{1tx} - kw_{3t}), & (6.1.4)\\
\rho c\theta_{3tt} = \theta_{3xxt} + \theta_{3xx} - \alpha T_0 \phi_{2tx}, & (6.1.5)
\end{cases}
$$

together with initial conditions

$$
\begin{cases}
w_1(x,0) = u_0(x),\ w_{1t}(x,0) = v_0(x),\ \phi_2(x,0) = \phi_0(x),\\
\phi_{2t}(x,0) = \psi_0(x),\ w_3(x,0) = w_0(x),\ w_{3t}(x,0) = \varphi_0(x),\\
\theta_1(x,0) = \theta_0(x),\ \theta_3(x,0) = \xi_0(x),\ \theta_{3t}(x,0) = \eta_0(x)
\end{cases}
\tag{6.1.6}
$$

and boundary conditions

$$
w_1(x,t) = w_{3x}(x,t) = \phi_2(x,t) = \theta_1(x,t) = \theta_3(x,t) = 0, \quad \text{for } x = 0, 1, \tag{6.1.7}
$$

where w_1, w_3, ϕ_2 are the longitudinal, vertical and shear angle displacement, θ_1, θ_3 are the temperature deviations from the T_0 along the longitudinal and vertical directions, $E, G, \rho, I, m, k, h, c$ are positive constants for the elastic and thermal material properties. We adopt the results in this chapter from [157].

Y. Qin and Z. Ma, *Global Well-posedness and Asymptotic Behavior of the Solutions to Non-classical Thermo(visco)elastic Models*,
DOI 10.1007/978-981-10-1714-8_6

6.2 Case of Equal Wave Speeds: E = G

Here we state and prove a decay result in the case of equal wave speeds propagation. Define the state spaces

$$\mathcal{H} = H_0^1 \times H_*^1 \times H_0^1 \times H_0^1 \times (L^2)^5,$$

where

$$H_*^1 = \left\{ f \in H^1(0,1) \,\middle|\, \int_0^1 f(x) = 0 \right\}.$$

The associated energy term is given by

$$\begin{aligned} E(t) = \frac{1}{2}\int_0^1 \Big(& [Eh(w_{1x} - kw_3)^2 + Gh(\phi_2 + w_{3x} + kw_1)^2 + EI\phi_{2x}^2] \\ & + [\rho h(w_{1t}^2 + w_{3t}^2) + \rho I \phi_{2t}^2] + \frac{\rho c}{T_0}(\theta_1^2 + \theta_{3t}^2 + \theta_{3x}^2)\Big) dx. \end{aligned} \tag{6.2.1}$$

By a straightforward calculation, we have

$$\frac{dE(t)}{dt} = -\frac{1}{T_0}\Big(\| \theta_{1x} \|^2 + \| \theta_{3xt} \|^2 \Big) \le 0. \tag{6.2.2}$$

From the semigroup theory ([151, 174]), we have the following existence and regularity result. For the explicit proofs, we refer the reader to [118] or Chap. 4 of this book.

Lemma 6.2.1 *Let* $(u_0(x), w_0(x), \varphi_0(x), \xi_0(x), v_0(x), \phi_0(x), \psi_0(x), \theta_0(x), \eta_0(x)) \in \mathcal{H}$. *Then problem (6.1.1)–(6.1.5) has a unique global weak solution* $(w_1, w_3, \phi_2, \theta_1, \theta_3)$ *verifying*

$$\begin{aligned} & w_3(x,t) \in C(\mathbb{R}^+, H_*^1(0,1)) \cap C^1(\mathbb{R}^+, L^2(0,1)), \\ & (w_1(x,t), \phi_2(x,t), \theta_1(x,t), \theta_3(x,t)) \in C(\mathbb{R}^+, H_0^1(0,1)) \cap C^1(\mathbb{R}^+, L^2(0,1)). \end{aligned}$$

We now state our main stability result.

Theorem 6.2.1 *Suppose that* $E = G$ *and* $(u_0(x), w_0(x), \varphi_0(x), \xi_0(x), v_0(x), \phi_0(x), \psi_0(x), \theta_0(x), \eta_0(x)) \in \mathcal{H}$, *then the energy* $E(t)$ *decays exponentially as time tends to* $+\infty$, *that is, there exist two positive constants* C *and* μ, *independent of the initial data and* t, *such that for all* $t > 0$,

$$E(t) \le CE(0)e^{-\mu t}. \tag{6.2.3}$$

The proof of Theorem 6.2.1 will be completed through several lemmas.

Let

$$I_1 = \int_0^1 (\rho I \phi_{2t}\phi_2 + \rho h w_{3t} f)\,dx, \tag{6.2.4}$$

where f is the solution of

$$-f_{xx} = \phi_{2x},\ f(0) = f(1) = 0. \tag{6.2.5}$$

Lemma 6.2.2 *Let $(w_1, w_3, \phi_2, \theta_1, \theta_3)$ be a solution of problem (6.1.1)–(6.1.5), then we have, for any $\varepsilon_1 > 0$,*

$$\begin{aligned}\frac{dI_1(t)}{dt} \le & -\frac{EI}{2} \parallel \phi_{2x} \parallel^2 + \rho I \parallel \phi_{2t} \parallel^2 + \varepsilon_1\Big(\parallel w_{3t} \parallel^2 + \parallel (w_{1x} - kw_3) \parallel^2 \Big)\\ & + C(\varepsilon_1)\Big(\parallel \theta_{3xt} \parallel^2 + \parallel \theta_{1x} \parallel^2 + \parallel \phi_{2t} \parallel^2 \Big).\end{aligned} \tag{6.2.6}$$

Proof Multiplying (6.1.2) by f, (6.1.3) by ϕ_2 respectively, and adding the results, we arrive at

$$\begin{aligned}\frac{dI_1(t)}{dt} = & -EI \parallel \phi_{2x} \parallel^2 + \rho I \parallel \phi_{2t} \parallel^2 - \int_0^1 \alpha\theta_{3xt}\phi_2 dx\\ & - kEh\int_0^1 (w_{1x} - kw_3) f dx - k\alpha \int_0^1 \theta_1 f dx + \rho h \int_0^1 w_{3t} f_t dx.\end{aligned} \tag{6.2.7}$$

By using the inequalities

$$\int_0^1 f_x^2 dx \le \int_0^1 \phi_2^2 dx \le \int_0^1 \phi_{2x}^2 dx,$$
$$\int_0^1 f_t^2 dx \le \int_0^1 f_{tx}^2 dx \le \int_0^1 \phi_{2t}^2 dx,$$

and Young's inequality, the assertion of the lemma follows. □

Let

$$I_2 = \rho c\rho h \int_0^1 \left(\int_0^x \theta_1 dy\right) w_{1t} dx. \tag{6.2.8}$$

Lemma 6.2.3 *Let $(w_1, w_3, \phi_2, \theta_1, \theta_3)$ be a solution of problem (6.1.1)–(6.1.5), then we have, for any $\varepsilon_2 > 0$,*

$$\frac{dI_2(t)}{dt} \le \frac{-\alpha\rho h T_0}{2}\int_0^1 w_{1t}^2dx + C(\varepsilon_2)\Big(\parallel \theta_{1x} \parallel^2 + \parallel w_{3t} \parallel^2 \Big)$$
$$+\varepsilon_2\Big(\parallel w_{1x} - kw_3 \parallel^2 + \parallel \phi_2 + w_{3x} + kw_1 \parallel^2 \Big). \qquad (6.2.9)$$

Proof Using Eqs. (6.1.4) and (6.1.1), we get

$$\begin{aligned}
\frac{dI_2(t)}{dt} &= \rho c\rho h\int_0^1\Big(\int_0^x \theta_{1t}dy\Big)w_{1t}dx + \rho c\rho h\int_0^1\Big(\int_0^x \theta_{1}dy\Big)w_{1tt}dx\\
&= \rho h\int_0^1\Big(\int_0^x \theta_{1xx} - \alpha T_0(w_{1tx} - kw_{3t})dy\Big)w_{1t}dx\\
&+\int_0^1\Big(\int_0^x \theta_1 dy\Big)\Big((Eh(w_{1x} - kw_3) - \alpha\theta_1)_x - KGh(\phi_2 + w_{3x} + kw_1)\Big)dx\\
&= \rho h\int_0^1 \theta_{1x}w_{1t}dx - \rho h\alpha T_0\int_0^1 w_{1t}^2dx + \rho hk\int_0^1\Big(\int_0^x w_{3t}dy\Big)w_{1t}dx\\
&+ \rho hEh\int_0^1\Big(\theta_{1x}w_1 + k\theta_1 w_3 + \alpha\theta_1^2\Big)dx\\
&- \rho ckGh\int_0^1\Big(\int_0^x \theta_1 dy\Big)(\phi_2 + w_{3x} + kw_1)dx.
\end{aligned}$$

The assertion of the lemma then follows by using Young's and Poincaré's inequalities. □

Let

$$I_3 = \rho^2 cI\int_0^1\Big(\int_0^x \theta_{3t}dy\Big)\phi_{2t}dx. \qquad (6.2.10)$$

Lemma 6.2.4 *Let $(w_1, w_3, \phi_2, \theta_1, \theta_3)$ be a solution of problem (6.1.1)–(6.1.5), then we have, for any $\varepsilon_3 > 0$,*

$$\frac{dI_3(t)}{dt} \le -\frac{\alpha\rho I T_0}{2} \parallel \phi_{2t} \parallel^2 + C(\varepsilon_3)(\parallel \theta_{3x} \parallel^2 + \parallel \theta_{3xt} \parallel^2)$$
$$+ \varepsilon_3 \parallel \phi_{2x} \parallel^2 + \varepsilon_3 \parallel \phi_2 + w_{3x} + kw_1 \parallel^2 . \qquad (6.2.11)$$

Proof Using Eqs. (6.1.3) and (6.1.5), we have

$$\begin{aligned}
\frac{dI_3}{dt} &= \rho c\rho I\int_0^1\Big(\int_0^x \theta_{3tt}dy\Big)\phi_{2t}dx + \rho c\rho I\int_0^1\Big(\int_0^x \theta_{3t}dy\Big)\phi_{2tt}dx\\
&= \rho I\int_0^1\int_0^x\Big(\theta_{3xxt} + \theta_{3xx} - \alpha T_0\phi_{2xt}\Big)dy\phi_{2t}dx\\
&+\rho c\int_0^1\Big(\int_0^x \theta_{3t}dy\Big)\Big(EI\phi_{2xx} - Gh(\phi_2 + w_{3x} + kw_1) - \alpha\theta_{3x}\Big)dx
\end{aligned}$$

$$= \rho I \int_0^1 (\theta_{3xt} + \theta_{3x})\phi_{2t} dx - \alpha I T_0 \int_0^1 \phi_{2t}^2 dx + \rho c E I \int_0^1 \theta_{3t}\phi_{2x} dx$$
$$-\rho c G h \int_0^1 \left(\int_0^x \theta_{3t} dy\right)(\phi_2 + w_{3x} + k w_1) dx - \alpha \rho c \int_0^1 \theta_{3t}^2 dx.$$

Then using Young's and Poincaré's inequalities, we can obtain the assertion. □

Next, we set

$$I_4 = h\rho I \int_0^1 \phi_{2t}(\phi_2 + w_{3x} + k w_1) dx + h\rho I \int_0^1 \phi_{2x} w_{3t} dx. \tag{6.2.12}$$

Lemma 6.2.5 *Let $(w_1, w_3, \phi_2, \theta_1, \theta_3)$ be a solution of problem (6.1.1)–(6.1.5), then we have, for any $\varepsilon_4 > 0$,*

$$\frac{dI_4(t)}{dt} \leq -\frac{Gh^2}{2}\int_0^1 (\phi_2 + w_{3x} + k w_1)^2 dx + C(\varepsilon_4)\Big(\| \theta_{3xt} \|^2 + \| \theta_{1x} \|^2 \Big)$$
$$+ \frac{kh\rho I}{2}\Big(\| \phi_{2t} \|^2 + \| w_{1t} \|^2 \Big) + C(\varepsilon_4) \| \phi_{2x} \|^2 + \varepsilon_4 \| w_{1x} - k w_3 \|^2 . \tag{6.2.13}$$

Proof Let $J_1 = I \int_0^1 \phi_{2t}(\phi_2 + w_{3x} + k w_1) dx$, $J_2 = h\rho I \int_0^1 \phi_{2x} w_{3t} dx$, then using Eqs. (6.1.2) and (6.1.3), we have

$$\frac{dJ_1(t)}{dt} = \rho I \int_0^1 \phi_{2tt}(\phi_2 + w_{3x} + k w_1) dx + h\rho I \int_0^1 \phi_{2t}(\phi_2 + w_{3x} + k w_1)_t dx$$
$$= hEI \int_0^1 \phi_{2xx}(\phi_2 + w_{3x} + k w_1) dx - Gh^2 \int_0^1 (\phi_2 + w_{3x} + k w_1)^2 dx$$
$$- \alpha h \int_0^1 \theta_{3xt}(\phi_2 + w_{3x} + k w_1) dx + h\rho I \int_0^1 \phi_{2t}^2 dx$$
$$+ h\rho I \int_0^1 \phi_{2t}(w_{3x} + k w_1)_t dx,$$

$$\frac{dJ_2(t)}{dt} = I\rho h \int_0^1 \phi_{2xt} w_{3t} dx + I\rho h \int_0^1 \phi_{2x} w_{3tt} dx$$
$$= -I\rho h \int_0^1 \phi_{2t} w_{3xt} dx + IGh \int_0^1 \phi_{2x}(\phi_2 + w_{3x} + k w_1)_x dx$$
$$+ IkEh \int_0^1 \phi_{2x}(w_{1x} - k w_3) dx - \alpha I k \int_0^1 \phi_{2x}\theta_1 dx.$$

Noticing that $E = G$, we get

$$\begin{aligned} I_4'(t) &= \frac{dJ_1}{dt} + \frac{dJ_2}{dt} \\ &= -Gh^2 \int_0^1 (\phi_2 + w_{3x} + kw_1)^2 dx - \alpha h \int_0^1 \theta_{3x}(\phi_2 + w_{3x} + kw_1)dx \\ &\quad + h\rho I \int_0^1 \phi_{2t}^2 dx + khI\rho \int_0^1 \phi_{2t} w_{1t} dx + IkEh \int_0^1 \phi_{2x}(w_{1x} - kw_3)dx \\ &\quad - \alpha Ik \int_0^1 \phi_{2x}\theta_1 dx. \end{aligned}$$

Then using Young's inequality, we can obtain the assertion. □

Set

$$I_5 = -h\rho \int_0^1 w_{3t}(w_{1x} - kw_3)dx - h\rho \int_0^1 w_{1t}(\phi_2 + w_{3x} + kw_1)dx. \quad (6.2.14)$$

Lemma 6.2.6 *Let $(w_1, w_3, \phi_2, \theta_1, \theta_3)$ be a solution of problem (6.1.1)–(6.1.5), then we have, for any $\varepsilon_5 > 0$,*

$$\begin{aligned} \frac{dI_5(t)}{dt} &\le -\frac{kEh}{2} \| (w_{1x} - kw_3) \|^2 - \frac{\rho h}{2} \| w_{1t} \|^2 + k\rho h \| w_{3t} \|^2 \\ &\quad + \frac{\rho h}{2} \| \phi_{2t} \|^2 + C(\varepsilon_5) \| \theta_{1x} \|^2 + (kGh + \varepsilon_5) \| (\phi_2 + w_{3x} + kw_1) \|^2 . \end{aligned} \quad (6.2.15)$$

Proof Let $H_1 = -h\rho \int_0^1 w_{3t}(w_{1x} - kw_3)dx$, $H_2 = -h\rho \int_0^1 w_{1t}(\phi_2 + w_{3x} + kw_1)dx$, then using Eqs. (6.1.1) and (6.1.2), we have

$$\begin{aligned} \frac{dH_1}{dt} &= -Gh \int_0^1 (\phi_2 + w_{3x} + kw_1)_x (w_{1x} - kw_3)dx - kEh \int_0^1 (w_{1x} - kw_3)^2 dx \\ &\quad + \alpha k \int_0^1 \theta_1 (w_{1x} - kw_3)dx + k\rho h \int_0^1 w_{3t}^2 - \rho h \int_0^1 w_{3t} w_{1xt} dx, \end{aligned}$$

$$\begin{aligned} \frac{dH_2}{dt} &= -Eh \int_0^1 (w_{1x} - kw_3)_x (\phi_2 + w_{3x} + kw_1)dx + \alpha \int_0^1 \theta_{1x}(\phi_2 + w_{3x} + kw_1)dx \\ &\quad + kGh \int_0^1 (\phi_2 + w_{3x} + kw_1)^2 dx - \rho h \int_0^1 w_{1t}^2 dx - \rho h \int_0^1 w_{1t}\phi_{2t} dx \\ &\quad + \rho h \int_0^1 w_{1tx} w_{3t} dx. \end{aligned}$$

Then, noticing $E = G$ again, from the above two equalities and Young's inequality, we can obtain the assertion. □

Next, we set

$$I_6 = -\rho h \int_0^1 w_{3t} w_3 dx - \rho h \int_0^1 w_{1t} w_1 dx. \tag{6.2.16}$$

Lemma 6.2.7 *Let $(w_1, w_3, \phi_2, \theta_1, \theta_3)$ be a solution of problem (6.1.1)–(6.1.5), then we have,*

$$\frac{dI_6(t)}{dt} \le -\rho h\Big(\| w_{3t} \|^2 + \| w_{1t} \|^2 \Big) + C \| \theta_{1x} \|^2 + C \| \phi_{2x} \|^2 . \tag{6.2.17}$$

Proof Using Eqs. (6.1.1) and (6.1.2), we have

$$\begin{aligned} I_6'(t) = & -\rho h \int_0^1 w_{3t}^2 dx - \rho h \int_0^1 w_{1t}^2 dx + Eh \int_0^1 (w_{1x} - kw_3)^2 dx \\ & + Gh \int_0^1 (\phi_2 + w_{3x} + kw_1)(w_{3x} + kw_1) dx - \alpha \int_0^1 \theta_1 (w_{1x} - kw_3) dx. \end{aligned} \tag{6.2.18}$$

Noticing (6.2.1) and (6.2.2), we have that there exists a constant $C > 0$ such that

$$-\alpha \int_0^1 \theta_1 (w_{1x} - kw_3) dx \le C \| \theta_{1x} \|^2 + Eh \| w_{1x} - kw_3 \|^2 . \tag{6.2.19}$$

Similarly,

$$\begin{aligned} & Gh \int_0^1 (\phi_2 + w_{3x} + kw_1)(w_{3x} + kw_1) dx \\ & = Gh \| \phi_2 + w_{3x} + kw_1 \|^2 - Gh \int_0^1 (\phi_2 + w_{3x} + kw_1) \phi_2 dx \\ & \le C \| \phi_{2x} \|^2 . \end{aligned} \tag{6.2.20}$$

Then inserting (6.2.19) and (6.2.20) into (6.2.18), the assertion of the lemma follows. □

Now, we set

$$I_7 = \rho c \int_0^1 \theta_{3t} \theta_3 dx + \frac{1}{2} \| \theta_{3x} \|^2 + \alpha T_0 \int_0^1 \phi_2 \theta_{3x} dx. \tag{6.2.21}$$

Lemma 6.2.8 *Let $(w_1, w_3, \phi_2, \theta_1, \theta_3)$ be a solution of problem (6.1.1)–(6.1.5), then we have, for any $\varepsilon_7 > 0$,*

$$\frac{dI_7(t)}{dt} \le -\parallel \theta_{3x} \parallel^2 + C(\varepsilon_7) \parallel \theta_{3xt} \parallel^2 + \varepsilon_7 \parallel \phi_{2x} \parallel^2 . \tag{6.2.22}$$

Proof Using (6.1.5), we have

$$\frac{dI_7(t)}{dt} = -\parallel \theta_{3x} \parallel^2 - \alpha T_0 \int_0^1 \phi_2 \theta_{3xt} dx.$$

Then using Young's and Poincaré's inequalities, we can obtain the assertion. □

Now, let $N, N_1, N_2, N_3, N_4, N_5, N_6, N_7 > 0$, and define the Lyapunov functional $\mathcal{F}(t)$ as follows

$$\mathcal{F}(t) = NE + N_1 I_1 + N_2 I_2 + N_3 I_3 + N_4 I_4 + N_5 I_5 + N_6 I_6 + N_7 I_7. \tag{6.2.23}$$

Using (6.2.2), (6.2.6), (6.2.9), (6.2.11), (6.2.13), (6.2.15), (6.2.17) and (6.2.22), we have

$$\begin{aligned} \frac{d\mathcal{F}(t)}{dt} &\le \Upsilon_1 \parallel \theta_{1x} \parallel^2 + \Upsilon_2 \parallel \theta_{3xt} \parallel^2 + \Upsilon_3 \parallel \phi_{2x} \parallel^2 + \Upsilon_4 \parallel w_{1t} \parallel^2 + \Upsilon_5 \parallel \phi_{2t} \parallel^2 \\ &+ \Upsilon_6 \parallel \phi_2 + w_{3x} + kw_1 \parallel^2 + \Upsilon_7 \parallel w_{1x} - kw_3 \parallel^2 + \Upsilon_8 \parallel w_{3t} \parallel^2 + \Upsilon_9 \parallel \theta_{3x} \parallel^2, \end{aligned} \tag{6.2.24}$$

where

$$\begin{cases} \Upsilon_1 = -\dfrac{N}{T_0} + C(\varepsilon_1)N_1 + N_2 C(\varepsilon_2) + N_4 C(\varepsilon_4) + N_5 C(\varepsilon_5) + C, \\ \Upsilon_2 = -\dfrac{N}{T_0} + C(\varepsilon_1)N_1 + N_3 C(\varepsilon_3) + N_4 C(\varepsilon_4) + N_7 C(\varepsilon_7), \\ \Upsilon_3 = -\dfrac{N_1 EI}{2} + \varepsilon_3 N_3 + C(\varepsilon_4) N_4 + C N_6, \\ \Upsilon_4 = -\dfrac{\alpha\rho h T_0 N_2}{2} + \dfrac{kh\rho I N_4}{2} - \dfrac{\rho h N_5}{2} - \rho h N_6, \\ \Upsilon_5 = -\dfrac{\alpha\rho h T_0 N_3}{2} + \dfrac{kh\rho I N_4}{2} + N_1 \rho I + N_1 C(\varepsilon_1) + \dfrac{\rho h N_5}{2}, \\ \Upsilon_6 = -\dfrac{G h^2 N_4}{2} + kGhN_5 + N_5\varepsilon_5 + N_3\varepsilon_3 + N_2\varepsilon_2, \\ \Upsilon_7 = -\dfrac{kEhN_5}{2} + N_4\varepsilon_4 + N_1\varepsilon_1 + N_2\varepsilon_2, \\ \Upsilon_8 = -N_6\rho h + N_5 k\rho h + C(\varepsilon_2)N_2 + N_1\varepsilon_1, \\ \Upsilon_9 = -N_7 + N_3 C(\varepsilon_3). \end{cases}$$

We choose our constants very carefully and properly so that there exists a constant $\omega > 0$, (6.2.24) takes the form

$$\frac{d\mathcal{F}(t)}{dt} \leq -\omega\Big(\parallel \theta_{1x} \parallel^2 + \parallel \theta_{3x} \parallel^2 + \parallel \theta_{3xt} \parallel^2 + \parallel \phi_{2x} \parallel^2 + \parallel w_{1t} \parallel^2$$
$$+ \parallel \phi_{2t} \parallel^2 + \parallel \phi_2 + w_{3x} + kw_1 \parallel^2 + \parallel w_{1x} - kw_3 \parallel^2 + \parallel w_{3t} \parallel^2 \Big). \tag{6.2.25}$$

We are now ready to prove Theorem 6.2.1. □

Proof of Theorem 6.2.1 Firstly, from the definition of $\mathcal{F}(t)$, we know that $\mathcal{F}(t)$ is equivalent to $E(t)$, i.e.,

$$\mathcal{F}(t) \sim E(t), \tag{6.2.26}$$

which, from (6.2.25) and (6.2.26), leads to

$$\frac{d}{dt}\mathcal{F}(t) \leq -\mu\mathcal{F}(t) \tag{6.2.27}$$

with a constant $\mu > 0$.
Integrating (6.2.27) over $(0, t)$, and using (6.2.26), leads to (6.2.3), we complete the proof of Theorem 6.2.1. □

6.3 Bibliographic Comments

From this seemingly complicated system (6.1.1)–(6.1.5), very interesting special cases can be obtained. In particular, the isothermal system is exactly the system obtained by Bresse [21] in 1856. The Bresse system, equations (6.1.1)–(6.1.3) with θ_1, θ_3 removed, is more general than the well-known Timoshenko system where the longitudinal displacement w_1 is not considered. If both θ_1 and w_1 are neglected, the Bresse thermoelastic system can be simplified to the following Timoshenko thermoelastic system

$$\begin{cases} \rho h w_{3tt} = Gh(\phi_2 + w_{3x})_x, & (6.3.1) \\ \rho I \phi_{2tt} = EI\phi_{2xx} - Gh(\phi_2 + w_{3x}) - \alpha\theta_{3tx}, & (6.3.2) \\ \rho c\theta_{3tt} = \theta_{3xxt} + \theta_{3xx} - \alpha T_0 \phi_{2tx}, & (6.3.3) \end{cases}$$

which was studied by Messaoudi and Said-Houari [175]. For the boundary conditions

$$w_3(x, t) = \phi_2(x, t) = \theta_{3x}(x, t) = 0, \quad \text{at} \quad x = 0, l, \tag{6.3.4}$$

they obtained the exponential stability for the thermoelastic Timoshenko system (6.1.1)-(6.1.3) when $E = G$. We refer the reader to the references [4, 80, 81, 149, 174, 191, 259] for the Timoshenko system with other kinds of damping mechanisms such as viscous damping, viscoelastic damping of Boltzmann type acting on the motion equation of w_3 or ϕ_2. Recently, Liu and Rao [151] considered a similar system, and used the semigroup method to show that the exponential decay rate is preserved when the wave speed of the vertical displacement coincides with the wave speed of longitudinal displacement or of the shear angle displacement. Otherwise, only a polynomial type decay rate can be obtained, their main tools are the frequency domain characterization of exponential decay obtained by Prüss [205] and Huang [118], and of polynomial decay obtained recently by Liu and Rao [150]. For other related works of this model, we refer the reader to the bibliographic comments in Sect. 5.5.

Chapter 7
Global Existence and Exponential Stabilization for the Higher-Dimensional Linear Thermoelastic System of Type III

7.1 Introduction

In this chapter, we shall consider the global existence and exponential stabilization of the linear thermoelastic model of type III with memory effects

$$\begin{cases} u_{tt} - \mu\Delta u - (\lambda+\mu)\nabla divu \\ \quad +\mu g * \Delta u + (\lambda+\mu) g * \nabla divu + \alpha\nabla\vartheta = 0, \quad (x,t) \in \Omega \times (0,+\infty), & (7.1.1) \\ \vartheta_{tt} - \Delta\theta_t - \Delta\vartheta + \beta \mathrm{div} u_{tt} = 0, \quad (x,t) \in \Omega \times (0,+\infty), & (7.1.2) \end{cases}$$

where the sign "$*$" denotes the convolution product in time, which is defined by

$$g * v(t) = \int_{-\infty}^{t} g(t-s)v(x,s)\,ds$$

with the initial data

$$u(x,0) = u_0(x), \quad u_t(x,0) = u_1(x), \quad \vartheta(x,0) = \vartheta_0(x), \quad x \in \Omega, \quad (7.1.3)$$

$$u(x,0) - u(x,-s) = w_0(x,s), \quad (x,s) \in \Omega \times (0,+\infty) \quad (7.1.4)$$

and boundary conditions

$$u = 0, \quad \vartheta = 0, \quad (x,t) \in \Gamma \times (0,+\infty). \quad (7.1.5)$$

System (7.1.1)–(7.1.2) is a model for a linear viscoelastic body Ω of the Boltzmann type with thermal damping. Here the body Ω is a bounded domain in $\mathbb{R}^n$ with smooth boundary $\Gamma = \partial\Omega$ (say C^2) and is assumed to be linear, homogeneous, and isotropic. Function $u(x,t) = (u_1(x,t), u_2(x,t), \cdots, u_n(x,t))$, and $\vartheta(x,t)$ represent displacement vector and temperature derivations, respectively, from the natural state of the reference configuration at position x and time t. Constants $\lambda, \mu > 0$ are

Y. Qin and Z. Ma, *Global Well-posedness and Asymptotic Behavior of the Solutions to Non-classical Thermo(visco)elastic Models*,
DOI 10.1007/978-981-10-1714-8_7

Lamé's constants and $\alpha, \beta > 0$ the coupling parameters; $g(t)$ denotes the relaxation function, $w_0(x, s)$ is a specified "history", and $u_0(x), u_1(x), \vartheta_0(x)$ are initial data. The notations $\Delta, \ \nabla, \ div$ denote the Laplace, gradient, and divergence operators in the space variables, respectively. The terms $\mu g * \Delta u$ and $(\lambda + \mu) g * \nabla \mathrm{div} u$ in (7.1.1) denote the memory effects.

The basic conditions on the relaxation function $g(t)$ are standard (see, e.g., Dafermos [47, 48] and Liu [148])

(H_1) $g \in C^1\Big([0, +\infty)\Big) \cap L^1(0, +\infty)$;
(H_2) $g(t) \geq 0, \quad g'(t) \leq 0, \quad t > 0$;
(H_3) $\kappa = 1 - \int_0^{+\infty} g(t)\, dt > 0$.

Condition (H_3) simply states that the static modulus of elasticity is positive. This restriction is quite natural. In addition, conditions (H_1) and (H_2) imply

$$g(+\infty) = \lim_{t \to +\infty} g(t) = 0. \tag{7.1.6}$$

In the sequel, we denote by $\|\cdot\|$ the norm of $L^2(\Omega)$. In order to simplify the notation, as in [285], we introduce the new variable

$$\theta(x, t) = \int_0^t \vartheta(x, s) ds + \chi(x) \tag{7.1.7}$$

with $\chi(x)$ to be defined below, instead of the temperature ϑ, of course, according to (7.1.7), we have

$$\theta_t = \vartheta. \tag{7.1.8}$$

Integrating (7.1.2) with respect to t, we deduce that

$$\theta_{tt} - \Delta\theta_t - \Delta\theta + \mathrm{div} u_t + \Delta\chi - \vartheta_1 + \Delta\vartheta_0 - \mathrm{div} u_1 = 0. \tag{7.1.9}$$

Therefore, by choosing $\chi \in H_0^1(\Omega)$ such that

$$\begin{cases} -\Delta\chi = \Delta\vartheta_0 - \mathrm{div} u_1 - \vartheta_1, \ in\ \Omega, \\ \chi = 0 \qquad\qquad on\ \partial\Omega, \end{cases} \tag{7.1.10}$$

system (7.1.1)–(7.1.5) reduces to the following equations

$$\begin{cases} u_{tt} - \mu\Delta u - (\lambda + \mu)\nabla div u \\ \qquad + \mu g * \Delta + (\lambda + \mu) g * \nabla div u + \alpha\nabla\theta_t = 0, \\ \theta_{tt} - \Delta\theta_t - \Delta\theta + \beta \mathrm{div} u_t = 0 \end{cases} \tag{7.1.11}$$

with the initial data

$$u(x,0) = u_0(x), \quad u_t(x,0) = u_1(x), \quad \theta(x,0) = \theta_0(x), \quad \theta_t(x,0) = \theta_1(x), \quad x \in \Omega. \tag{7.1.12}$$

Thus we shall analyze system (7.1.11)–(7.1.12).

The energy $E(u, \theta, t)$ of system (7.1.11)–(7.1.12) is defined by

$$\begin{aligned} E(u,\theta,t) &= \frac{1}{2}\Big\{ \| u_t(t) \|^2 + \frac{\alpha}{\beta}(\| \theta_t(t) \|^2 + \| \nabla\theta \|^2)\Big\} + \frac{\mu\kappa}{2} \| \nabla u(t) \|^2 \\ &+ \frac{(\lambda+\mu)\kappa}{2} \| divu(t) \|^2 + \frac{\mu}{2}\int_{-\infty}^{t} g(t-s) \| \nabla u(t) - \nabla u(s) \|^2 \, ds \\ &+ \frac{\lambda+\mu}{2}\int_{-\infty}^{t} g(t-s) \| divu(t) - divu(s) \|^2 \, ds. \end{aligned}$$

Here we have used the notation

$$\|v\|^2 = \sum_{i=1}^{n} \|v_i\|^2, \quad for \;\; v = (v_1, v_2, \cdots, v_n).$$

By a straightforward calculation, we have

$$\begin{aligned} \frac{dE(t)}{dt} &= \frac{\mu}{2}\int_{-\infty}^{t} g_t(t-s) \| \nabla u(t) - \nabla u(s) \|^2 \, ds \\ &+ \frac{\lambda+\mu}{2}\int_{-\infty}^{t} g_t(t-s) \| divu(t) - divu(s) \|^2 \, ds - \frac{\alpha}{\beta} \| \nabla\theta_t(t) \|^2 \le 0 \end{aligned} \tag{7.1.13}$$

which indicates that the energy $E(u, \theta, t)$ decreases on $(0, +\infty)$. However, the most interesting question is whether the energy decays exponentially as $t \to +\infty$. Namely, are there positive constants M, ω such that

$$E(u,\theta,t) \le Me^{-\omega t} E(u,\theta,0), \quad for \;\; all \; t \ge 0. \tag{7.1.14}$$

In order to design a boundary velocity feedback, we put

$$\Gamma_1 = \{x \in \Gamma \; : m(x) \cdot \nu(x) \le 0\} \tag{7.1.15}$$

and

$$\Gamma_2 = \{x \in \Gamma \; : m(x) \cdot \nu(x) \ge 0\} \tag{7.1.16}$$

where

$$m(x) = x - x^0 = (x_1 - x_1^0, x_2 - x_2^0, \cdots, x_n - x_n^0) \tag{7.1.17}$$

for some $x^0 \in \mathbb{R}^n$, $\nu = (\nu_1, \nu_2, \cdots, \nu_n)$ denotes the unit outward normal on Γ to Ω and

$$m \cdot \nu = m(x) \cdot \nu(x) = \sum_{i=1}^{n} (x_i - x_i^0)\nu_i. \tag{7.1.18}$$

A part of Γ, Γ_1 is assumed either to be empty or to have a nonempty interior relative to Γ. Note (7.1.15) and (7.1.16) imply that the domain Ω is simply connected and star-shaped with respect to $x^0 \in \Omega$ or $\Omega = \Omega_1 - \overline{\Omega}_2$. Both Ω_1 and Ω_2 are star-shaped with respect to x^0.

Thus the boundary velocity feedback can be given as follows

$$\begin{cases} \theta = 0, \quad (x,t) \in \Gamma \times (0, +\infty), & (7.1.19) \\ u = 0, \quad (x,t) \in \Gamma_1 \times (0, +\infty), & (7.1.20) \\ \mu \dfrac{\partial}{\partial \nu}(u - g * u) + (\lambda + \mu)\nu div(u - g * u) & \\ \quad + am \cdot \nu(u - g * u) + m \cdot \nu u_t = 0, (x,t) \in \Gamma_2 \times (0, +\infty) & (7.1.21) \end{cases}$$

where $a = a(x)$ is a given non-negative function on Γ_2 with

$$a(x) \in C^1(\Gamma_2). \tag{7.1.22}$$

It is clear that if $m(x) \cdot \nu(x) \geq \eta$ on Γ_2 for some constant $\eta > 0$, then $a(x)m(x) \cdot \nu(x)$ can be any non-negative function as we can take $a(x) = f(x)/(m(x) \cdot \nu(x))$, $f(x)$ being any non-negative function. Note that there is no velocity feedback on the part of Γ_2 where $m(x) \cdot \nu(x) = 0$. We shall prove that the energy of the thermoelastic system with boundary velocity feedback (7.1.19)–(7.1.21) decays to zero exponentially as $t \to +\infty$ if $A = \max_{x \in \Gamma_2} a(x)$ is small enough and the following conditions hold

$$\Gamma_1 \neq \emptyset \quad or \quad a(x) \not\equiv 0 \tag{7.1.23}$$

and

$$\overline{\Gamma}_1 \cap \overline{\Gamma}_2 = \emptyset. \tag{7.1.24}$$

We adopt the results in this chapter from [220].

7.2 Main Results

Let $a(x)$ be the non-negative function given in (7.1.21) satisfying (7.1.22) and $\nu = (\nu_1, \cdots, \nu_n)$ the unit normal on Γ directed towards the exterior of Ω. Suppose that Γ_1 and Γ_2 are given by (7.1.15) and (7.1.16), respectively, and Γ_1 either is empty or has a non-empty interior relative to Γ. Assume that (7.1.23)–(7.1.24) hold.

We further introduce some function space. Set

$$H^1_{\Gamma_1}(\Omega) = \{u \in H^1(\Omega) \ : u = 0 \ on \ \Gamma_1\}. \tag{7.2.1}$$

We define the norm of $H^1_{\Gamma_1}(\Omega)$ by

$$\|u\|_{H^1_{\Gamma_1}(\Omega)} = \left(\frac{\mu}{2}\|\nabla u\|^2 + \frac{\lambda+\mu}{2}\|divu\|^2 + \frac{1}{2}\int_{\Gamma_2} am \cdot \nu |u|^2 \, d\Gamma\right)^{\frac{1}{2}}. \tag{7.2.2}$$

Under conditions (7.1.23)–(7.1.24), this norm is equivalent to the usual one induced by $H^1(\Omega)$. Let the "*history space*" $L^2(g, (0, +\infty), (H^1_{\Gamma_1}(\Omega))^n)$ consist of $((H^1_{\Gamma_1}(\Omega))^n)$-valued functions w on $(0, +\infty)$ for which

$$\|w\|^2_{L^2(g, (0,+\infty), (H^1_{\Gamma_1}(\Omega))^n)} = \int_0^{+\infty} g(s)\|w(s)\|^2_{(H^1_{\Gamma_1}(\Omega))^n} \, ds < +\infty. \tag{7.2.3}$$

Put

$$\mathcal{H} = (H^1_{\Gamma_1}(\Omega))^n \times (L^2(\Omega))^n \times H^1_0(\Omega) \times L^2(\Omega) \times L^2(g, (0, +\infty), (H^1_{\Gamma_1}(\Omega))^n) \tag{7.2.4}$$

with the energy norm

$$\begin{aligned} &\|(u, v, \theta, \theta_t, w)\|_{\mathcal{H}} \\ &= \left\{\kappa\|u\|^2_{(H^1_{\Gamma_1}(\Omega))^n} + \frac{1}{2}\Big(\|v\|^2 + \frac{\alpha}{\beta}(\|\theta_t\|^2 + \|\nabla\theta\|^2)\Big)\right. \\ &\quad \left. + \int_0^{+\infty} g(s)\|w(s)\|^2_{(H^1_{\Gamma_1}(\Omega))^n} \, ds\right\}^{\frac{1}{2}} \end{aligned} \tag{7.2.5}$$

where κ denotes the positive constant in (H_3), that is,

$$\kappa = 1 - \int_0^{+\infty} g(t) \, dt > 0. \tag{7.2.6}$$

We further introduce three constants as follows. Set

$$R_0 = \max_{x\in\bar{\Omega}} | \, m(x) \, | = \max_{x\in\bar{\Omega}} | \sum_{k=1}^{n} (x_k - x_k^0)^2 \, |^{\frac{1}{2}} \tag{7.2.7}$$

where $m(x)$ is given by (7.1.17). Let γ be the smallest positive constant such that

$$\int_{\Gamma_2} | u |^2 d\Gamma \leq \gamma^2 \| u \|_{H^1_{\Gamma_1}(\Omega)}, \quad for \;\; all \;\; u \in H^1_{\Gamma_1}(\Omega). \tag{7.2.8}$$

For each time t, we may regard u and θ as elements of function spaces. Accordingly, we suppress their argument $x \in \mathbb{R}^n$ from the notation.

Now consider the following system with a boundary velocity feedback

$$\begin{cases} u_{tt} - \mu\Delta u - (\lambda+\mu)\nabla \mathrm{div} u \\ \quad +\mu g * \Delta u + (\lambda+\mu) g * \nabla \mathrm{div} u + \alpha\nabla\theta_t = 0, \quad (x,t) \in \Omega \times (0,+\infty), & (7.2.9) \\ \theta_{tt} - \Delta\theta_t - \Delta\theta + \beta \mathrm{div} u_t = 0, \quad (x,t) \in \Omega \times (0,+\infty), & (7.2.10) \\ \theta = 0, \quad (x,t) \in \Gamma \times (0,+\infty), & (7.2.11) \\ u = 0, \quad (x,t) \in \Gamma_1 \times (0,+\infty), & (7.2.12) \\ \mu \frac{\partial}{\partial \nu}(u - g * u) + (\lambda+\mu)\nu \mathrm{div}(u - g * u) \\ \quad +am \cdot \nu(u - g * u) + m \cdot \nu u_t = 0, \quad (x,t) \in \Gamma_2 \times (0,+\infty), & (7.2.13) \\ u(x,0) = u_0(x), \ u_t(x,0) = u_1(x), \ \theta(x,0) = \theta_0(x), \quad x \in \Omega, & (7.2.14) \\ u(x,0) - u(x,-s) = w_0(x,s), \quad (x,t) \in \Omega \times (0,+\infty). & (7.2.15) \end{cases}$$

Let

$$v(x,t) = u_t(x,t), \quad w^t(x,s) = u(x,t) - u(x,t-s). \tag{7.2.16}$$

We shall prove that problem (7.2.9)–(7.2.15) is globally well-posed in Sect. 7.3. In fact, we shall prove that the system generates a strongly continuous semigroup $S(t)$ of contractions on $\mathcal{H}$. Furthermore, under the assumptions (7.1.23)–(7.1.24), we ensure that the solution u of problem (7.2.9)–(7.215) has the sufficient regularity and satisfies

$$u \in C([0,+\infty), (H^2(\Omega))^n). \tag{7.2.17}$$

The energy $E(u,\theta,t)$ of problem (7.2.9)–(7.2.15) is defined by:

$$\begin{aligned} E(u,\theta,t) &= \| \Big(u(t), u_t(t), \theta(t), \theta_t(t), u(t) - u(t-s)\Big) \|^2_{\mathcal{H}} \\ &= \kappa \| u(t) \|^2_{(H^1_{\Gamma_1}(\Omega))^n} + \frac{1}{2}\Big(\| u_t(t) \|^2 + \frac{\alpha}{\beta}(\| \theta_t(t) \|^2 + \| \nabla\theta(t) \|^2)\Big) \\ &\quad + \int_0^{+\infty} g(s) \| u(t) - u(t-s) \|^2_{(H^1_{\Gamma_1}(\Omega))^n} ds \\ &= \kappa \| u(t) \|^2_{(H^1_{\Gamma_1}(\Omega))^n} + \frac{1}{2}\Big(\| u_t(t) \|^2 + \frac{\alpha}{\beta}(\| \theta_t(t) \|^2 + \| \nabla\theta(t) \|^2)\Big) \\ &\quad + \int_{-\infty}^{t} g(t-s) \| u(t) - u(s) \|^2_{(H^1_{\Gamma_1}(\Omega))^n} ds. \end{aligned} \tag{7.2.18}$$

By a straightforward calculation, we have

$$\begin{aligned}\frac{dE}{dt} &= \int_{-\infty}^{t} g_t(t-s) \parallel u(t)-u(s) \parallel^2_{(H^1_{\Gamma_1}(\Omega))^n} ds \\ &- \int_{\Gamma_2} m\cdot\nu \mid u_t(t) \mid^2 d\Gamma - \frac{\alpha}{\beta} \parallel \nabla\theta_t(t) \parallel^2 . \end{aligned} \tag{7.2.19}$$

Therefore the energy $E(u, \theta, t)$ decreases in $(0, +\infty)$, and moreover, we have the following exponential decay rate, which is our main result of this chapter.

Theorem 7.2.1 *Let Γ_1 and Γ_2 be given by (7.1.15) and (7.1.16), respectively, satisfying (7.1.22)–(7.1.24). Let (7.1.23)–(7.1.24) hold. Suppose that the relaxation function g satisfies $(H_1)-(H_3)$ and the following condition:*
(H_4) there exists a constant $K>0$ such that for all $t\in(0,+\infty)$

$$-G(t) = \int_t^{+\infty} g(s)ds \le Kg(t). \tag{7.2.20}$$

If the function $a(x)$ satisfies

$$2\left(\frac{2a^2R_0^2}{\mu} + (2-n)a\right) R_0\gamma^2(1+(1-\kappa)^2) < \kappa, \quad \text{for } n \le 2, \tag{7.2.21}$$

$$a \le \frac{(n-2)\mu}{2R_0^2}, \quad \text{for } n \ge 3, \tag{7.2.22}$$

then there are positive constants M, ω, independent of $(u_0, u_1, \theta_0, \theta_1, w_0)$, such that

$$E(u,\theta,t) \le ME(u,\theta,0)e^{-\omega t}, \quad for \;\; all\; t\ge 0 \tag{7.2.23}$$

for all solutions to problem (7.2.9)–(7.2.15) with $(u_0, u_1, \theta_0, \theta_1, w_0) \in \mathcal{H}$. Furthermore, the positive constants M, ω can be explicitly given by

$$M = (\rho(T))^{-1}, \omega = -T^{-1}\ln\rho(T), \tag{7.2.24}$$

$$\rho(T) = \frac{exp(\delta T/(1+\delta C_1))(1-\delta C_1)\kappa + 2\delta C_5 K}{\kappa(1-\delta C_1)}, \tag{7.2.25}$$

$$T = 1 + \frac{\delta(1+\delta C_1)}{4}\ln\frac{2(1+\delta C_1)}{1-\delta C_1}, \tag{7.2.26}$$

$$\delta = \min\left\{\frac{\kappa}{4KC_5 + C_1\kappa}, \frac{1}{C_3}, \frac{1}{C_4}, \frac{4\alpha(\lambda+\mu)\kappa}{2(\lambda+\mu)\kappa(2\beta C_2 + \alpha\lambda_0^2 + \alpha\lambda_1^2) + \alpha^2\lambda_1^2\beta}\right\}, \quad (7.2.27)$$

$$C_1 = \max\{c_1, c_2, c_3\}, \quad (7.2.28)$$

with

$$\begin{aligned} c_1 &= 2R_0(2-\kappa) + \mid n-2 \mid (2-\kappa) + 1, \\ c_2 &= \frac{1}{\kappa}\left(\frac{2R_0(3-2\kappa)}{\mu} + \mid n-2 \mid \lambda_0^2(\frac{3}{2}-\kappa) + \frac{\lambda_0^2}{2}\right), \\ c_3 &= \frac{4R_0}{\mu} + \lambda_0^2, \end{aligned}$$

$$C_2 = \frac{\alpha^2(8R_0^2(2-\kappa)/\mu + \lambda_0^2(2-\kappa) \mid n-2 \mid + \lambda_0^2)}{4\varepsilon}, \quad (7.2.29)$$

$$C_3 = \frac{2R_0^2}{\mu} + \frac{\gamma^2 R_0((2-\kappa) \mid n-2 \mid + 1)}{4\varepsilon} + 1, \quad (7.2.30)$$

$$C_4 = \frac{g(0)(8R_0^2 + \mu \mid n-2 \mid \lambda_0^2)}{4\varepsilon\mu}, \quad (7.2.31)$$

$$C_5 = \begin{cases} 1 + \varepsilon + 2 \mid n-2 \mid \varepsilon + 2K(A)(1-\kappa)R_0\gamma^2, & n \le 2, \\ 1 + \varepsilon + 2 \mid n-2 \mid \varepsilon, & n \ge 3, \end{cases} \quad (7.2.32)$$

$$K(A) = \frac{2A^2R_0^2}{\mu} - (n-2)A, \quad A = \max_{x\in\Gamma_2} a(x), \quad (7.2.33)$$

$$0 < \varepsilon = \begin{cases} \min\left\{\frac{1}{2(1+|n-2|)}, \frac{\kappa - 2K(A)R_0\gamma^2(1+(1-\kappa)^2)}{3+2|n-2|+(1+2|n-2|)(1-\kappa)}\right\}, & n \le 2, \\ \min\left\{\frac{1}{2(1+|n-2|)}, \frac{\kappa}{3+2|n-2|+(1+2|n-2|)(1-\kappa)}\right\}, & n \ge 3, \end{cases} \quad (7.2.34)$$

where λ_0 is the best constant in Poincaré's inequality.

7.3 Global Well-Posedness

We shall briefly prove global existence result by using the semigroup approach in this section.

In order to use the theory of semigroups, we write problem (7.2.9)–(7.2.15) as follows:

$$\begin{cases} u_{tt} - \kappa\mu\Delta u - \kappa(\lambda+\mu)\nabla \mathrm{div} u + \alpha\nabla\theta_t - \mu\int_0^{+\infty} g(s)\Delta w^t(x,s)\,ds \\ -(\lambda+\mu)\int_0^{+\infty} g(s)\nabla \mathrm{div} w^t(x,s)\,ds = 0, \quad (x,t)\in\Omega\times(0,+\infty), & (7.3.1)\\ \theta_{tt} - \Delta\theta_t - \Delta\theta + \beta \mathrm{div} u_t = 0, \quad (x,t)\in\Omega\times(0,+\infty), & (7.3.2)\\ w^t(x,s) = u(x,t) - u(x,t-s), \quad (x,t,s)\in\Omega\times(0,+\infty)\times(0,+\infty), & (7.3.3)\\ \theta = 0, \quad (x,t)\in\Gamma\times(0,+\infty), & (7.3.4)\\ u = 0, \quad (x,t)\in\Gamma_1\times(0,+\infty), & (7.3.5)\\ \kappa\mu\frac{\partial u}{\partial\nu} + \kappa(\lambda+\mu)\nu \mathrm{div} u + \mu\int_0^{+\infty} g(s)\frac{\partial w^t(x,s)}{\partial\nu}\,ds + \kappa a m\cdot\nu u \\ \quad +(\lambda+\mu)\nu\int_0^{+\infty} g(s)\mathrm{div} w^t(x,s)\,ds + a m\cdot\nu\int_0^{+\infty} g(s) w^t(x,s)\,ds \\ \quad + m\cdot\nu u_t = 0, (x,t)\in\Gamma_2\times(0,+\infty), & (7.3.6)\\ u(x,0)=u_0(x),\ u_t(x,0)=u_1(x),\ \theta(x,0)=\theta_0(x), \quad x\in\Omega, & (7.3.7)\\ w^0(x,s) = w_0(s), \quad (x,t)\in\Omega\times(0,+\infty). & (7.3.8) \end{cases}$$

We define a linear unbounded operator A on $\mathcal{H}$ by

$$A(u,v,\theta,\theta_t,w) = (v, B(u,w) - \alpha\nabla\theta_t, \theta_t, \Delta\theta_t + \Delta\theta - \beta \mathrm{div} v, v - w_s) \quad (7.3.9)$$

where $w_s = \frac{\partial w}{\partial s}$ and

$$B(u,w) = \kappa\mu\Delta u + \kappa(\lambda+\mu)\nabla \mathrm{div} u + \mu\int_0^{+\infty} g(s)\Delta w^t(x,s)\,ds + (\lambda+\mu)\int_0^{+\infty} g(s)\nabla \mathrm{div} w^t(x,s)\,ds.$$

Set

$$\Phi = (u, v, \theta, \theta_t, w).$$

Then problem (7.3.1)–(7.3.8) can be formulated as an abstract Cauchy problem

$$\frac{d\Phi}{dt} = A\Phi,$$

on the Hilbert space $\mathcal{H}$ for an initial condition $\Phi(0) = (u_0, u_1, \theta_0, \theta_1, w_0)$. The domain of A is given by:

$$\begin{aligned} D(A) = \Big\{ &(u, v, \theta, w) \in \mathcal{H} \ : \theta \in H_0^1(\Omega), \theta_t \in H_0^1(\Omega), \ \theta + \theta_t \in H^2(\Omega) \cap H_0^1(\Omega), \\ & v \in (H_{\Gamma_1}^1(\Omega))^n, \ \kappa u + \int_0^{+\infty} g(s) w^t(x, s)\, ds \in (H^2(\Omega) \cap H_{\Gamma_1}^1(\Omega))^n, \\ & w^t(x, s) \in H^1(g, (0, +\infty), (H_{\Gamma_1}^1(\Omega))^n), w^t(x, 0) = 0, \\ & \kappa\mu \frac{\partial u}{\partial \nu} + \kappa(\lambda + \mu)\nu \text{div} u + \kappa a m \cdot \nu u \\ & + \mu \int_0^{+\infty} g(s) \frac{\partial w^t(x, s)}{\partial \nu}\, ds + (\lambda + \mu)\nu \int_0^{+\infty} g(s) \text{div} w^t(x, s)\, ds \\ & + a m \cdot \nu \int_0^{\infty} g(s) w^t(x, s)\, ds + m \cdot \nu v = 0 \quad on\ \Gamma_2 \Big\} \end{aligned} \tag{7.3.10}$$

where

$$H^1(g, (0, +\infty), (H_{\Gamma_1}^1(\Omega))^n) = \{ w^t \ : w^t, \ w_s^t \in L^2(g, (0, +\infty), (H_{\Gamma_1}^1(\Omega))^n) \}.$$

It is clear that $D(A)$ is dense in $\mathcal{H}$.

To prove that A is dissipative, we need the following lemma.

Lemma 7.3.1 *If the function $f : [0, +\infty) \to \mathbb{R}$ is uniformly continuous and is in $L^1(0, +\infty)$, then*

$$\lim_{t \to +\infty} f(t) = 0.$$

Proof See, e.g., Qin [210]. □

Lemma 7.3.2 *Suppose that the relaxation function g satisfies (H_1) and (H_2). If $w^t \in H^1(g, (0, +\infty), (H_{\Gamma_1}^1(\Omega))^n)$ and $w^t(x, 0) = 0$, then*

$$g'(s) \parallel w^t(x, s) \parallel^2_{(H_{\Gamma_1}^1(\Omega))^n} \in L^1(0, +\infty)$$

and

$$\lim_{s \to +\infty} \left(g(s) \parallel w^t(x, s) \parallel^2_{(H_{\Gamma_1}^1(\Omega))^n} \right) = 0.$$

Proof For $w^t \in H^1(g, (0, +\infty), (H_{\Gamma_1}^1(\Omega))^n)$ and $w^t(x, 0) = 0$, we have

$$\lim_{s \to 0} \left(g(s) \parallel w^t(x, s) \parallel^2_{H_{\Gamma_1}^1(\Omega))^n} \right) = \lim_{s \to 0} \left(g(0) \parallel w^t(x, 0) \parallel^2_{(H_{\Gamma_1}^1(\Omega))^n} \right) = 0.$$

It therefore follows that

$$
\begin{aligned}
&2\int_0^t g(s)\Big(w_s^t(x,s), w^t(x,s)\Big)_{(H^1_{\Gamma_1}(\Omega))^n} ds \\
&= \int_0^t g(s)\frac{\partial}{\partial s}\left(\| w^t(x,s)\|^2_{(H^1_{\Gamma_1}(\Omega))^n}\right) ds \\
&= g(t)\| w^t(x,s)\|^2_{(H^1_{\Gamma_1}(\Omega))^n} - \int_0^t g'(s)\| w^t(x,s)\|^2_{(H^1_{\Gamma_1}(\Omega))^n} ds,
\end{aligned}
$$

which implies that for all $t \geq 0$

$$
\begin{aligned}
\int_0^t g'(s)\| w^t(x,s)\|^2_{(H^1_{\Gamma_1}(\Omega))^n} ds &\leq 2\left(\int_0^{+\infty} g(s)\| w_s^t(x,s)\|^2_{(H^1_{\Gamma_1}(\Omega))^n} ds\right)^{1/2} \\
&\times \left(\int_0^{+\infty} g(s)\| w^t(x,s)\|^2_{(H^1_{\Gamma_1}(\Omega))^n} ds\right)^{1/2}.
\end{aligned}
$$

Thus

$$
g'(s)\| w_s^t(x,s)\|^2_{(H^1_{\Gamma_1}(\Omega))^n} \in L^1(0,+\infty). \tag{7.3.11}
$$

On the other hand, for any $0 \leq s_1 < s_2 < +\infty$, we have

$$
\begin{aligned}
&g(s_1)\| w_s^t(s_1)\|^2_{(H^1_{\Gamma_1}(\Omega))^n} - g(s_2)\| w_s^t(s_2)\|^2_{(H^1_{\Gamma_1}(\Omega))^n} \\
&= \int_{s_1}^{s_2} \frac{d}{ds}\left(g(s)\| w_s^t(x,s)\|^2_{(H^1_{\Gamma_1}(\Omega))^n}\right) ds \\
&= \int_{s_1}^{s_2} g'(s)\| w_s^t(x,s)\|^2_{(H^1_{\Gamma_1}(\Omega))^n} ds + 2\int_{s_1}^{s_2} g(s)(w_s^t(x,s), w^t(x,s))_{(H^1_{\Gamma_1}(\Omega))^n} ds,
\end{aligned}
$$

which, combined with (7.3.11), implies that $g(s)\| w_s^t(x,s)\|^2_{(H^1_{\Gamma_1}(\Omega))^n}$ is uniformly continuous on $[0,+\infty)$. Hence Lemma 7.3.1 gives

$$
\lim_{s\to+\infty}\left(g(s)\| w_s^t(x,s)\|^2_{(H^1_{\Gamma_1}(\Omega))^n}\right) = 0.
$$

□

Lemma 7.3.3 *Let Γ_1 and Γ_2 be given by (7.1.15)–(7.1.16), respectively, satisfying (7.1.22)–(7.1.24). Suppose that the relaxation function g satisfies $(H_1) - (H_3)$. Then the operator A is dissipative and closed.*

Proof By a straightforward calculation, it follows from Lemma 7.3.2 that

$$
\begin{aligned}
&\langle A(u, v, \theta, \theta_t, w^t), (u, v, \theta, \theta_t, w^t)\rangle_{\mathcal{H}} \\
&= \kappa(v, u)_{(H^1_{\Gamma_1}(\Omega))^n} + \frac{1}{2}\left(B(u, w^t) - \alpha\nabla\theta_t, v\right) + \frac{\alpha}{2\beta}(\nabla\theta_t, \nabla\theta) \\
&\quad + \frac{\alpha}{2\beta}(\Delta\theta_t + \Delta\theta - \mathrm{div}v, \theta_t) + (v - w^t_s, w^t)_{L^2(g,(0,+\infty),(H^1_{\Gamma_1}(\Omega))^n)} \\
&= -\frac{1}{2}\int_{\Gamma_2} m\cdot v \mid v\mid^2 d\Gamma - \frac{\alpha}{2\beta}\parallel \nabla\theta_t \parallel^2 - g(s)\parallel w^t(x, s)\parallel^2_{(H^1_{\Gamma_1}(\Omega))^n}\Big|_0^{+\infty} \\
&\quad + \int_0^{+\infty} g'(s)\parallel w^t(x, s)\parallel^2_{(H^1_{\Gamma_1}(\Omega))^n} ds \\
&\leq 0.
\end{aligned}
$$

Thus A is dissipative. □

To prove that A is closed, let $(u_n, v_n, \theta_n, \theta_{nt}, w^t_n) \in D(A)$ be such that

$$
(u_n, v_n, \theta_n, \theta_{nt}, w^t_n) \to (u, v, \theta, \theta_t, w^t) \quad \text{in } \mathcal{H}
$$

and

$$
A(u_n, v_n, \theta_n, \theta_{nt}, w^t_n) \to (a, b, c, d, e) \quad \text{in } \mathcal{H}.
$$

Then we have

$$
\begin{cases}
u_n \to u \quad \text{strongly} \quad \text{in } (H^1_{\Gamma_1}(\Omega))^n, & (7.3.12)\\
v_n \to v \quad \text{strongly} \quad \text{in } (L^2(\Omega))^n, & (7.3.13)\\
\theta_n \to \theta \quad \text{strongly} \quad \text{in } H^1_0(\Omega), & (7.3.14)\\
\theta_{nt} \to \theta_t \quad \text{strongly} \quad \text{in } L^2(\Omega), & (7.3.15)\\
w^t_n \to w^t \quad \text{strongly} \quad \text{in } L^2(g, (0, +\infty), (H^1_{\Gamma_1}(\Omega))^n), & (7.3.16)
\end{cases}
$$

and

$$
\begin{cases}
v_n \to a \quad \text{strongly} \quad \text{in } (H^1_{\Gamma_1}(\Omega))^n, & (7.3.17)\\
B(u_n, w^t_n) - \alpha\nabla\theta_{nt} \to b \quad \text{strongly} \quad \text{in } (L^2(\Omega))^n, & (7.3.18)\\
\theta_{nt} \to c \quad \text{strongly} \quad \text{in } H^1_0(\Omega), & (7.3.19)\\
\Delta\theta_{nt} + \Delta\theta_n - \beta\mathrm{div}v_n \to d \quad \text{strongly} \quad \text{in } L^2(\Omega), & (7.3.20)\\
v_n - w^t_{ns} \to e \quad \text{strongly} \quad \text{in } L^2(g, (0, +\infty), (H^1_{\Gamma_1}(\Omega))^n). & (7.3.21)
\end{cases}
$$

By (7.3.13) and (7.3.17), we deduce

$$
v_n \to v \quad \text{strongly} \quad \text{in } (H^1_{\Gamma_1}(\Omega))^n, \tag{7.3.22}
$$

and

$$v = a \in (H^1_{\Gamma_1}(\Omega))^n. \tag{7.3.23}$$

By (7.3.15) and (7.3.19), we deduce

$$\theta_{nt} \to \theta_t \quad \text{strongly} \quad \text{in } H^1_0(\Omega), \tag{7.3.24}$$

and

$$\theta_t = c \in H^1_0(\Omega). \tag{7.3.25}$$

By (7.3.20) and (7.3.22), we deduce

$$\Delta\theta_{nt} + \Delta\theta_n \to d + \beta \mathrm{div} v \quad \text{strongly} \quad \text{in } L^2(\Omega), \tag{7.3.26}$$

and consequently, it follows from (7.3.14), (7.3.24) that

$$\theta_{nt} + \theta_n \to \theta_t + \theta \quad \text{strongly} \quad \text{in } H^2(\Omega) \cap H^1_0(\Omega), \tag{7.3.27}$$

since Δ is an isomorphism from $H^2(\Omega) \cap H^1_0(\Omega)$ onto $L^2(\Omega)$. It therefore follows from (7.3.20) and (7.3.27) that

$$d = \Delta\theta_t + \Delta\theta - \beta \mathrm{div} v, \quad \theta_t + \theta \in H^2(\Omega) \cap H^1_0(\Omega). \tag{7.3.28}$$

By (7.3.16), (7.3.21), and (7.3.22), we deduce

$$w^t_n \to w^t \quad \text{strongly} \quad \text{in } H^1(g, (0, +\infty), (H^1_{\Gamma_1}(\Omega))^n) \tag{7.3.29}$$

and

$$e = v - w^t_s, \quad w^t \in H^1(g, (0, +\infty), (H^1_{\Gamma_1}(\Omega))^n), \; w^t(x, 0) = 0. \tag{7.3.30}$$

In addition, it follows from (7.3.12), (7.3.16), (7.3.24) that

$$B(u_n, w^t_n) - \alpha\nabla\theta_{nt} \to B(u, w^t) - \alpha\nabla\theta_t \tag{7.3.31}$$

in the sense of distribution. It therefore follows from (7.3.18) and (7.3.31) that

$$b = B(u, w^t) - \alpha\nabla\theta_t, \quad B(u, w^t) \in (L^2(\Omega))^n \tag{7.3.32}$$

and consequently,

$$\kappa u + \int_0^{+\infty} g(s) w^t(x, s) ds \in (H^2(\Omega) \cap H^1_{\Gamma_1}(\Omega))^n, \tag{7.3.33}$$

since $\mu\Delta + (\lambda + \mu)\nabla \text{div}$ is an isomorphism from $H^2(\Omega) \cap H^1_{\Gamma_1}(\Omega)$ onto $L^2(\Omega)$. Moreover, by (7.3.22)–(7.3.33) and the trace theorem, we deduce that:

$$\kappa\mu\frac{\partial u}{\partial \nu} + \kappa(\lambda + \mu)\nu div u + \mu \int_0^{+\infty} g(s)\frac{\partial w^t(x, s)}{\partial \nu} ds$$
$$+\kappa a m \cdot \nu u + (\lambda + \mu)\nu \int_0^{+\infty} g(s) div w^t(x, s)\, ds$$
$$+a m \cdot \nu \int_0^{+\infty} g(s) w^t(x, s)\, ds + m \cdot \nu v = 0 \quad \text{on} \quad \Gamma_2.$$

Thus from (7.2.3), (7.3.25), (7.3.28), (7.3.30), (7.3.32) and (7.3.33), it follows

$$A(u, v, \theta, \theta_t, w^t) = (a, b, c, d, e), \quad (u, v, \theta, \theta_t, w^t) \in D(A).$$

Hence, A is closed. □

Lemma 7.3.4 $0 \in \rho(A)$, *where* $\rho(A)$ *is the resolvent set of the operator* A.

Proof For any $G = (g_1, g_2, g_3, g_4, g_5) \in \mathcal{H}$, consider

$$A\Phi = G, \tag{7.3.34}$$

i.e.,

$$\begin{cases} v = g_1, & in \ \ (H^1_{\Gamma_1}(\Omega))^n, & (7.3.35) \\ B(u, w^t) - \alpha\nabla\theta_t = g_2, & in \ \ (L^2(\Omega))^n, & (7.3.36) \\ \theta_t = g_3, & in \ \ H^1_0(\Omega), & (7.3.37) \\ \Delta\theta_t + \Delta\theta - \beta div v = g_4, & in \ \ L^2(\Omega), & (7.3.38) \\ v - w^t_s = g_5, & in \ \ L^2(g, (0, +\infty), (H^1_{\Gamma_1}(\Omega))^n). & (7.3.39) \end{cases}$$

Inserting $v = g_1$ and $\theta_t = g_3$ obtained from (7.3.35), (7.3.37) into (7.3.38), we obtain

$$\Delta\theta = g_4 + \beta div g_1 - \Delta g_3 \in L^2(\Omega). \tag{7.3.40}$$

By the standard theory for the linear elliptic equations, we have a unique $\theta \in H^2(\Omega) \cap H^1_0(\Omega)$ satisfying (7.3.40).

We plug $v = g_1$ obtained from (7.3.35) into (7.3.39) to get

$$w^t_s = g_1 - g_5 \in L^2(g, (0, +\infty), (H^1_{\Gamma_1}(\Omega))^n). \tag{7.3.41}$$

Applying the standard theory for the linear elliptic equations again, we have a unique $w \in H^1(g, (0, +\infty), (H^1_{\Gamma_1}(\Omega))^n)$ satisfying (7.3.41). Then plugging θ and w just obtained from solving (7.3.40), (7.3.41), respectively, into (7.3.36) and applying the standard theory for the linear elliptic equations again yields the unique solvability of $u \in D(A)$ for (7.3.36), and such that $\kappa u + \int_0^{+\infty} g(s)w^t(x,s)\,ds \in (H^2(\Omega) \cap H^1_{\Gamma_1}(\Omega))^n$. Thus the unique solvability of (7.3.34) follows. It is clear from the regularity theory for the linear elliptic equations that $\|\Phi\|_{\mathcal{H}} \le K\|G\|_{\mathcal{H}}$ with K being a positive constant independent of Φ. Thus the proof is complete. □

From Lemmas 7.3.3–7.3.4 and Corollary 4.4 in [45], we have the following theorem.

Theorem 7.3.5 *Let Γ_1 and Γ_2 be given by (7.1.15)–(7.1.16), respectively, satisfying (7.1.22)–(7.1.24). Suppose that the relaxation function g satisfies $(H_1) - (H_3)$. Then*

(i) For every initial condition $(u_0, v_0, \theta_0, \theta_1, w^0) \in \mathcal{H}$, problem (7.3.1)–(7.3.8) has a unique global mild solution satisfying

$$(u, v, \theta, \theta_t, w^t) \in C([0, +\infty); \mathcal{H}).$$

Moreover, we have for every $t \in [0, +\infty)$,

$$\| (u(t), v(t), \theta(t), \theta_t(t), w^t(x,s)) \|_{\mathcal{H}} \le C \| (u_0, v_0, \theta_0, \theta_1, w^0) \|_{\mathcal{H}}. \tag{7.3.42}$$

(ii) For every initial condition $(u_0, v_0, \theta_0, \theta_1, w^0) \in D(A)$, problem (7.3.1)–(7.3.8) has a unique global classical solution satisfying

$$(u, v, \theta, \theta_t, w^t) \in C([0, +\infty); D(A)).$$

7.4 Proof of Main Results

It suffices to show that there exist positive constants $T > 0$ and $0 < \rho < 1$ such that for all $t \ge T$,

$$E(u, \theta, t) \le \rho E(u, \theta, 0). \tag{7.4.1}$$

However, the verification of (7.4.1) is generally not easy. Here we borrow the idea of the Lyapunov method to prove it, the Lyapunov functional to be constructed actually is a generalized energy functional which is closely related to the energy functional $E(u, \theta, t)$. Such similar Lyapunov functional was constructed for the wave equation, thermoelastic plate models and others. As we shall see, our Lyapunov functional does not exactly possess the properties that the Lyapunov functional defined for the nonlinear dynamical systems has. Therefore, we may call our Lyapunov functional a generalized Lyapunov functional.

In what follows, we assume the summation convention for repeated indices.

Let (u, θ) be the solution of problem (7.2.9)–(7.2.15) and δ any positive number. Similar Lyapunov functionals are constructed as

$$V(u,\theta,t)=E(u,\theta,t)+\delta\Big(F(u,\theta,t)+F_1(u,\theta,t)\Big), \tag{7.4.2}$$

where

$$F(u,\theta,t)=\int_\Omega\Big(2u_{it}m\cdot\nabla(u_i-g*u_i)+(n-2)u_{it}(u_i-g*u_i)+u_iu_{it}\Big)dx, \tag{7.4.3}$$

$$F_1(u,\theta,t)=\frac{\alpha}{2\beta}\int_\Omega\Big(\theta_t\theta+\frac{1}{2}\mid\nabla\theta\mid^2+\beta\text{div}u\cdot\theta\Big)dx. \tag{7.4.4}$$

As we have assumed the summation convention, equality (7.4.3) means that

$$F(u,\theta,t)=\sum_{i=1}^n\int_\Omega\Big(2u_{it}m\cdot\nabla(u_i-g*u_i)+(n-2)u_{it}(u_i-g*u_i)+u_iu_{it}\Big)dx.$$

In order to show that V is a generalized Lyapunov functional, we need to estimate $V'(u,\theta,t)$. We begin with the following lemma.

Lemma 7.4.1 *Let Γ_1 and Γ_2 be given by (7.1.15)–(7.1.16), respectively, satisfying (7.1.22)–(7.1.24). Suppose that the relaxation function g satisfies $(H_1)-(H_3)$. Then we have*

$$(1-\delta C_1)E(u,\theta,t)\le V(u,\theta,t)\le(1+\delta C_1)E(u,\theta,t), \tag{7.4.5}$$

for all solutions (u,θ) of problem (7.2.9)–(7.2.15), where the positive constant C_1, independent of (u,θ), is given by (7.2.28) and δ is given by (7.2.27).

Proof It is easy to see that

$$\begin{aligned}
&\Big|2\int_\Omega u_{it}(m\cdot\nabla(u_i-g*u_i))dx\Big|\\
&=\Big|2\int_\Omega u_{it}m\cdot\nabla u_idx-2\int_{-\infty}^t g(t-s)ds\int_\Omega u_{it}(t)\cdot\nabla u_i(s)dx\Big|\\
&\le R_0(\|u_{it}(t)\|^2+\|\nabla u_i(t)\|^2)\\
&\quad+R_0\int_{-\infty}^t g(t-s)(\|u_{it}(t)\|^2+\|\nabla u_i(s)\|^2)ds\\
&=R_0(2-\kappa)\|u_{it}(t)\|^2+R_0\|\nabla u_i(t)\|^2\\
&\quad+R_0\int_{-\infty}^t g(t-s)\|\nabla u_i(t)-\nabla u_i(s)-\nabla u_i(t)\|^2)ds\\
&\le R_0(2-\kappa)\|u_{it}(t)\|^2+R_0(3-2\kappa)\|\nabla u_i(t)\|^2\\
&\quad+2R_0\int_{-\infty}^t g(t-s)\|\nabla u_i(t)-\nabla u_i(s)\|^2)ds.
\end{aligned} \tag{7.4.6}$$

Let λ_0 be the best constant in Poincaré's inequality, namely, the smallest positive constant such that

$$\| u \| \leq \lambda_0 \| u \|_{H^1_{\Gamma_1}(\Omega)}, \quad for \ \ all \ u \in H^1_{\Gamma_1}(\Omega). \tag{7.4.7}$$

Using (7.4.7), we obtain

$$\begin{aligned}
&| \int_\Omega u_{it}(u_i - g * u_i)dx | \\
&= | \int_\Omega u_{it}u_i dx - \int_{-\infty}^t g(t-s)ds \int_\Omega u_{it}(t)u_i(s)dx | \\
&\leq \frac{1}{2} \| u_{it}(t) \|^2 + \frac{\lambda_0^2}{2} \| u_i(t) \|^2_{H^1_{\Gamma_1}(\Omega)} \\
&+ \frac{1}{2} \int_{-\infty}^t g(t-s)\Big(\| u_{it} \|^2 + \lambda_0^2 \| u_i(t) \|^2_{H^1_{\Gamma_1}(\Omega)} \Big) ds \\
&\leq \frac{2-\kappa}{2} \| u_{it}(t) \|^2 + \lambda_0^2 (\frac{3}{2} - \kappa) \| u_i(t) \|^2_{H^1_{\Gamma_1}(\Omega)} \\
&+ \lambda_0^2 \int_{-\infty}^t g(t-s) \| u_i(t) - u_i(s) \|^2_{H^1_{\Gamma_1}(\Omega)} ds
\end{aligned} \tag{7.4.8}$$

and

$$| \int_\Omega u_{it}u_i dx | \leq \frac{1}{2} \| u_{it}(t) \|^2 + \frac{\lambda_0^2}{2} \| u_i(t) \|^2_{H^1_{\Gamma_1}(\Omega)} . \tag{7.4.9}$$

Noting (7.2.18), we derive from (7.4.6), (7.4.8) and (7.4.9) that

$$| F(u, \theta, t) | \leq C_1 E(u, \theta, t)$$

which implies (7.4.5). □

Lemma 7.4.2 *Let* $g \in L^1(0, +\infty)$. *Set*

$$\begin{aligned}
H(u, \theta, t) = &\int_\Omega \Big(\frac{1}{2} | u_{it} |^2 + \mu \nabla u_i \nabla (u_i - g * u_i) + (\lambda + \mu) \mathrm{div} u \ \mathrm{div}(u - g * u) \Big) dx \\
&+ \int_{\Gamma_2} am \cdot \nu u_i (u_i - g * u_i) d\Gamma
\end{aligned}$$

and

$$G(t) = -\int_t^{+\infty} g(s)ds. \tag{7.4.10}$$

Then we have

$$H(u,\theta,t)=E(u,\theta,t)-\frac{\alpha}{2\beta}(\|\theta_t(t)\|^2+\|\nabla\theta(t)\|^2)+\kappa\|u(t)\|^2_{(H^1_{\Gamma_1}(\Omega))^n}$$
$$-\frac{d}{dt}\int_{-\infty}^t G(t-s)\|u(s)\|^2_{(H^1_{\Gamma_1}(\Omega))^n}ds \tag{7.4.11}$$

and

$$\int_{-\infty}^t g(t-s)\|u(s)\|^2_{(H^1_{\Gamma_1}(\Omega))^n}ds$$
$$=\frac{d}{dt}\int_{-\infty}^t G(t-s)\|u(s)\|^2_{(H^1_{\Gamma_1}(\Omega))^n}ds+(1-\kappa)\|u(t)\|^2_{(H^1_{\Gamma_1}(\Omega))^n}. \tag{7.4.12}$$

Proof By a straightforward calculation, we obtain

$$\int_{-\infty}^t g(t-s)\|\nabla u_i(t)-\nabla u_i(s)\|^2ds$$
$$=(1-\kappa)\|\nabla u_i(t)\|^2+\int_{-\infty}^t g(t-s)\|\nabla u_i(s)\|^2ds$$
$$-2\int_{-\infty}^t\int_\Omega g(t-s)\nabla u_i(t)\nabla u_i(s)dxds$$
$$=-(1+\kappa)\|\nabla u_i(t)\|^2+\int_{-\infty}^t g(t-s)\|\nabla u_i(s)\|^2ds$$
$$+2\int_\Omega\nabla u_i(t)\nabla(u_i(t)-g*u_i(t))dx \tag{7.4.13}$$

and

$$\int_{-\infty}^t g(t-s)\|\nabla u_i(s)\|^2ds$$
$$=\frac{d}{dt}\int_{-\infty}^t G(t-s)\|\nabla u_i(s)\|^2ds+(1-\kappa)\|\nabla u_i(t)\|^2. \tag{7.4.14}$$

Similarly, we have

$$\int_{-\infty}^t g(t-s)\|\mathrm{div}u(t)-\mathrm{div}u(s)\|^2ds$$
$$=-(1+\kappa)\|\mathrm{div}u(t)\|^2+\int_{-\infty}^t g(t-s)\|\mathrm{div}u(s)\|^2ds$$
$$+2\int_\Omega\mathrm{div}u(t)\mathrm{div}(u(t)-g*u(t))dx, \tag{7.4.15}$$

$$\int_{-\infty}^{t} g(t-s) \parallel \mathrm{div} u(s) \parallel^2 ds$$
$$= \frac{d}{dt}\int_{-\infty}^{t} G(t-s) \parallel \mathrm{div} u(s) \parallel^2 ds + (1-\kappa) \parallel \mathrm{div} u(t) \parallel^2, \tag{7.4.16}$$

$$\int_{-\infty}^{t} g(t-s) \int_{\Gamma_2} am\cdot\nu \mid u_i(t) - u_i(s) \mid^2 d\Gamma ds$$
$$= -(1+\kappa)\int_{\Gamma_2} am\cdot\nu \mid u_i(t) \mid^2 d\Gamma + \int_{-\infty}^{t} g(t-s)\int_{\Gamma_2} am\cdot\nu \mid u_i(s) \mid^2 d\Gamma ds$$
$$+ 2\int_{\Gamma_2} am\cdot\nu u_i(t)(u_i(t) - g * u_i(t)) d\Gamma, \tag{7.4.17}$$

$$\int_{-\infty}^{t} g(t-s) \int_{\Gamma_2} am\cdot\nu \mid u_i(s) \mid^2 d\Gamma ds$$
$$= \frac{d}{dt}\int_{-\infty}^{t} G(t-s)\int_{\Gamma_2} am\cdot\nu \mid u_i(s) \mid^2 d\Gamma ds + (1-\kappa)\int_{\Gamma_2} am\cdot\nu \mid u_i(t) \mid^2 d\Gamma. \tag{7.4.18}$$

Hence (7.4.11) and (7.4.12) follows from (7.4.13)–(7.4.18). □

Lemma 7.4.3 *Let Γ_1 and Γ_2 be given by (7.1.15)–(7.1.16), respectively, satisfying (7.1.22)–(7.1.24). Suppose that the relaxation function g satisfies $(H_1) - (H_3)$. Let (u, θ) be the solution of problem (7.2.9)–(7.2.15). If the function $a(x)$ satisfies (7.2.21) and (7.2.22), then we have*

$$\frac{dF}{dt}(u, \theta, t)$$
$$\leq -E(u, \theta, t) + (C_2 + \frac{\alpha\lambda_0^2}{2\beta}) \parallel \nabla\theta_t(t) \parallel^2 + \frac{\alpha}{2\beta} \parallel \nabla\theta(t) \parallel^2 + C_3\int_{\Gamma_2} m\cdot\nu \mid u_t \mid^2 d\Gamma$$
$$+ C_4\int_{-\infty}^{t} \mid g_t(t-s) \mid\parallel u(t) - u(s) \parallel^2_{(H^1_{\Gamma_1}(\Omega))^n} ds$$
$$+ C_5\frac{d}{dt}\int_{-\infty}^{t} G(t-s) \parallel u(s) \parallel^2_{(H^1_{\Gamma_1}(\Omega))^n} ds, \tag{7.4.19}$$

where positive constants C_2, C_3, C_4, C_5 are given by (7.2.29)–(7.2.32), respectively.

Proof By (7.4.3), we derive

$$\frac{dF}{dt}(u,\theta,t) = \int_\Omega 2u_{itt} m\cdot\nabla(u_i - g*u_i)dx + \int_\Omega 2u_{it} m\cdot\nabla(u_i - g*u_i)_t dx$$
$$+\int_\Omega (n-2)(u_i - g*u_i)u_{itt}dx + \int_\Omega (n-2)(u_i - g*u_i)_t u_{it}dx$$
$$+\int_\Omega u_i u_{itt}dx + \| u_{it} \|^2 . \tag{7.4.20}$$

We now estimate each term in (7.4.20) as follows. Since $u - g*u = 0$ on Γ_1, we get

$$\frac{\partial}{\partial x_k}(u_i - g*u_i) = \frac{\partial}{\partial \nu}(u_i - g*u_i)\nu_k \quad \text{on} \quad \Gamma_1. \tag{7.4.21}$$

Thus using (7.2.9) and boundary condition (7.2.13), we obtain

$$2\int_\Omega u_{itt} m\cdot\nabla(u_i - g*u_i)dx$$
$$= 2\int_\Omega \left(\mu\Delta(u_i - g*u_i) + (\lambda+\mu)\frac{\partial}{\partial x_i}(\mathrm{div}(u - g*u)) - \alpha\frac{\partial\theta_t}{\partial x_i}\right) \times m\cdot\nabla(u_i - g*u_i)dx$$
$$= \mu\int_\Gamma \left(2\frac{\partial}{\partial \nu}(u_i - g*u_i)m\cdot\nabla(u_i - g*u_i) - m\cdot\nu \mid \nabla(u_i - g*u_i)\mid^2\right)d\Gamma$$
$$+(n-2)\mu \| \nabla(u_i - g*u_i)\|^2$$
$$+(\lambda+\mu)\int_\Gamma 2\mathrm{div}(u - g*u)m_k\nu_i\frac{\partial}{\partial x_k}(u_i - g*u_i)d\Gamma$$
$$-(\lambda+\mu)\int_\Gamma m\cdot\nu \mid \mathrm{div}(u - g*u)\mid^2 d\Gamma$$
$$+(n-2)(\lambda+\mu)\| \mathrm{div}(u - g*u)\|^2 - 2\alpha\int_\Omega \left(m\cdot\nabla(u_i - g*u_i)\right)\frac{\partial\theta_t}{\partial x_i}dx$$
$$= \int_{\Gamma_1} m\cdot\nu\left(\mu \mid \frac{\partial}{\partial \nu}(u_i - g*u_i)\mid^2 + (\lambda+\mu)\mid \mathrm{div}(u - g*u)\mid^2\right)d\Gamma \quad (= I_1)$$
$$+2\int_{\Gamma_2}\left(\mu\frac{\partial}{\partial \nu}(u_i - g*u_i) + (\lambda+\mu)\nu_i\mathrm{div}(u - g*u)\right)m\cdot\nabla(u_i - g*u_i)d\Gamma$$
$$-\int_{\Gamma_2} m\cdot\nu\left(\mu \mid \nabla(u_i - g*u_i)\mid^2 + (\lambda+\mu)\mid \mathrm{div}(u - g*u)\mid^2\right)d\Gamma \quad (= I_3)$$
$$+(n-2)\left(\mu \| \nabla(u_i - g*u_i)\|^2 + (\lambda+\mu)\| \mathrm{div}(u - g*u)\|^2\right) \quad (= I_4)$$
$$-2\alpha\int_\Omega\left(m\cdot\nabla(u_i - g*u_i)\right)\frac{\partial\theta_t}{\partial x_i}dx \quad (= I_5)$$
$$= I_1 + I_3 + I_4 + I_5$$
$$-2\int_{\Gamma_2}(am\cdot\nu(u_i - g*u_i) + m\cdot\nu u_{it})m\cdot\nabla(u_i - g*u_i)d\Gamma \quad (= I_2). \tag{7.4.22}$$

Since $m\cdot\nu \le 0$ on Γ_1, we have

$$I_1 \le 0. \tag{7.4.23}$$

Since

$$I_2 \le \int_{\Gamma_2} m \cdot \nu \left(\frac{2a^2 R_0^2}{\mu} \mid u_i - g * u_i \mid^2 + \frac{2R_0^2}{\mu} \mid u_{it} \mid^2 + \mu \mid \nabla(u_i - g * u_i) \mid^2 \right) d\Gamma, \quad (7.4.24)$$

we have

$$I_2 + I_3 \le \int_{\Gamma_2} m \cdot \nu \left(\frac{2a^2 R_0^2}{\mu} \mid u_i - g * u_i \mid^2 + \frac{2R_0^2}{\mu} \mid u_{it} \mid^2 \right) d\Gamma. \quad (7.4.25)$$

In addition, we have for any $\varepsilon > 0$,

$$\begin{aligned} I_5 &\le \frac{\varepsilon\mu}{2} \parallel \nabla u_i(t) \parallel^2 + \frac{2\alpha^2 R_0^2}{\mu\varepsilon} \parallel \nabla\theta_t(t) \parallel^2 \\ &\quad + \int_{-\infty}^t g(t-s) \left(\frac{\varepsilon\mu}{2} \parallel \nabla u_i(s) \parallel^2 + \frac{2\alpha^2 R_0^2}{\mu\varepsilon} \parallel \nabla\theta_t(t) \parallel^2 \right) ds \\ &\le \varepsilon \parallel u_i(t) \parallel^2_{H^1_{\Gamma_1}(\Omega)} + \frac{2\alpha^2 R_0^2 (2-\kappa)}{\mu\varepsilon} \parallel \nabla\theta_t(t) \parallel^2 \\ &\quad + \varepsilon \int_{-\infty}^t g(t-s) \parallel u_i(s) \parallel^2_{H^1_{\Gamma_1}(\Omega)} ds. \end{aligned} \quad (7.4.26)$$

It therefore follows from (7.4.22)–(7.4.26) that

$$\begin{aligned} &2\int_\Omega u_{itt} m \cdot \nabla(u_i - g * u_i) dx \\ &\le \int_{\Gamma_2} m \cdot \nu \left(\frac{2a^2 R_0^2}{\mu} \mid u_i - g * u_i \mid^2 + \frac{2R_0^2}{\mu} \mid u_{it} \mid^2 \right) d\Gamma \\ &\quad + \varepsilon \parallel u_i(t) \parallel^2_{H^1_{\Gamma_1}(\Omega)} + \frac{2\alpha^2 R_0^2 (2-\kappa)}{\mu\varepsilon} \parallel \nabla\theta_t(t) \parallel^2 \\ &\quad + \varepsilon \int_{-\infty}^t g(t-s) \parallel u_i(s) \parallel^2_{H^1_{\Gamma_1}(\Omega)} ds \\ &\quad + (n-2)\Big(\mu \parallel \nabla(u_i(t) - g * u_i(t)) \parallel^2 + (\lambda+\mu) \parallel \mathrm{div}(u(t) - g * u(t)) \parallel^2 \Big). \end{aligned} \quad (7.4.27)$$

Since by (7.1.6),

$$-g(0) = \int_0^{+\infty} g_t(s) ds = \int_{-\infty}^t g_t(t-s) ds, \quad (7.4.28)$$

we have

$$
\begin{aligned}
&2\int_\Omega u_{it} m\cdot\nabla(u_i - g*u_i)_t dx\\
&= 2\int_\Omega u_{it} m_k\left(\frac{\partial u_{it}}{\partial x_k} - g(0)\frac{\partial u_i}{\partial x_k} - \int_{-\infty}^t g_t(t-s)\frac{\partial u_i(s)}{\partial x_k}ds\right)dx\\
&= 2\int_\Omega u_{it} m_k\Big(\int_{-\infty}^t g_t(t-s)(\frac{\partial u_i(t)}{\partial x_k} - \frac{\partial u_i(s)}{\partial x_k})ds\Big)dx\\
&\quad - n \parallel u_{it}(t)\parallel^2 + \int_{\Gamma_2} m\cdot\nu \mid u_{it}\mid^2 d\Gamma\\
&\le -n\parallel u_{it}(t)\parallel^2 + \int_{\Gamma_2} m\cdot\nu\mid u_{it}\mid^2 d\Gamma\\
&\quad + \int_{-\infty}^t \mid g_t(t-s)\mid \left(\frac{\varepsilon}{g(0)}\parallel u_{it}\parallel^2 + \frac{g(0)R_0^2}{\varepsilon}\parallel \nabla u_i(t) - \nabla u_i(s)\parallel^2\right)ds\\
&= (\varepsilon - n)\parallel u_{it}(t)\parallel^2 + \int_{\Gamma_2} m\cdot\nu\mid u_{it}\mid^2 d\Gamma\\
&\quad + \frac{2g(0)R_0^2}{\mu\varepsilon}\int_{-\infty}^t \mid g_t(t-s)\mid\parallel u_i(t) - u_i(s)\parallel^2_{H^1_{\Gamma_1}(\Omega)} ds. \qquad (7.4.29)
\end{aligned}
$$

Using (7.2.8), (7.2.9), (7.4.7), boundary condition (7.2.13) and Young's inequality, we deduce

$$
\begin{aligned}
&(n-2)\int_\Omega (u_i - g*u_i)u_{itt}dx\\
&= (n-2)\int_\Omega (u_i - g*u_i)\Big(\mu\Delta(u_i - g*u_i)\\
&\quad + (\lambda+\mu)\frac{\partial}{\partial x_i}(\mathrm{div}(u - g*u)) - \alpha\frac{\partial\theta_t}{\partial x_i}\Big)dx\\
&= (n-2)\int_{\Gamma_2}\Big(\mu\frac{\partial}{\partial\nu}(u_i - g*u_i) + (\lambda+\mu)\nu_i\mathrm{div}(u - g*u)\Big)(u_i - g*u_i)d\Gamma\\
&\quad - (n-2)\Big(\mu\parallel\nabla(u_i(t) - g*u_i(t))\parallel^2 + (\lambda+\mu)\parallel\mathrm{div}(u(t) - g*u(t))\parallel^2\Big)\\
&\quad - (n-2)\alpha\int_\Omega (u_i - g*u_i)\frac{\partial\theta_t}{\partial x_i}dx\\
&= -(n-2)\int_{\Gamma_2}\Big(am\cdot\nu(u_i - g*u_i) + m\cdot\nu u_{it}\Big)(u_i - g*u_i)d\Gamma\\
&\quad - (n-2)\Big(\mu\parallel\nabla(u_i(t) - g*u_i(t))\parallel^2 + (\lambda+\mu)\parallel\mathrm{div}(u(t) - g*u(t))\parallel^2\Big)\\
&\quad - (n-2)\alpha\int_\Omega\frac{\partial\theta_t}{\partial x_i}\Big(u_i - \int_{-\infty}^t g(t-s)u_i(s)ds\Big)dx
\end{aligned}
$$

$$
\begin{aligned}
&\leq -(n-2)\int_{\Gamma_2} am\cdot\nu \mid u_i - g*u_i\mid^2 d\Gamma \\
&\quad + \mid n-2\mid \int_{\Gamma_2} m\cdot\nu\Big(\frac{R_0\gamma^2}{4\varepsilon}\mid u_{it}\mid^2 + \frac{\varepsilon}{R_0\gamma^2}\mid u_i\mid^2\Big)d\Gamma \\
&\quad + \mid n-2\mid \int_{-\infty}^{t} g(t-s)ds\int_{\Gamma_2} m\cdot\nu\Big(\frac{R_0\gamma^2}{4\varepsilon}\mid u_{it}(t)\mid^2 + \frac{\varepsilon}{R_0\gamma^2}\mid u_i(s)\mid^2\Big)d\Gamma \\
&\quad - (n-2)\Big(\mu \parallel \nabla(u_i(t) - g*u_i(t))\parallel^2 + (\lambda+\mu)\parallel \operatorname{div}(u(t)-g*u(t))\parallel^2\Big) \\
&\quad + \mid n-2\mid \frac{\alpha^2\lambda_0^2}{4\varepsilon}\parallel \nabla\theta_t(t)\parallel^2 + \varepsilon\mid n-2\mid\parallel u_i(t)\parallel^2_{H^1_{\Gamma_1}(\Omega)} \\
&\quad + \mid n-2\mid \int_{-\infty}^{t} g(t-s)\Big(\frac{\alpha^2\lambda_0^2}{4\varepsilon}\parallel \nabla\theta_t(t)\parallel^2 + \frac{\varepsilon}{\lambda_0^2}\parallel u_i(s)\parallel^2\Big)ds \\
&\leq -(n-2)\int_{\Gamma_2} am\cdot\nu\mid u_i - g*u_i\mid^2 d\Gamma \\
&\quad + \frac{\mid n-2\mid \gamma^2 R_0(2-\kappa)}{4\varepsilon}\int_{\Gamma_2} m\cdot\nu\mid u_{it}\mid^2 d\Gamma + 2\varepsilon\mid n-2\mid\parallel u_i(t)\parallel^2_{H^1_{\Gamma_1}(\Omega)} \\
&\quad + \frac{\mid n-2\mid \alpha^2\lambda_0(2-\kappa)}{4\varepsilon}\parallel \nabla\theta_t(t)\parallel^2 \\
&\quad + 2\varepsilon\mid n-2\mid \int_{-\infty}^{t} g(t-s)\parallel u_i(t)\parallel^2_{H^1_{\Gamma_1}(\Omega)} ds \\
&\quad - (n-2)\Big(\mu\parallel \nabla(u_i(t)-g*u_i(t))\parallel^2 + (\lambda+\mu)\parallel \operatorname{div}(u(t)-g*u(t))\parallel^2\Big).
\end{aligned}
\tag{7.4.30}
$$

Using (7.4.7), (7.4.28) and Young's inequality, we obtain

$$
\begin{aligned}
&(n-2)\int_{\Omega} u_{it}(u_i - g*u_i)_t dx \\
&\quad = (n-2)\int_{\Omega} u_{it}\Big(u_{it} - g(0)u_i - \int_{-\infty}^{t} g_t(t-s)u_i(s)ds\Big)dx \\
&\quad = (n-2)\parallel u_{it}\parallel^2 + (n-2)\int_{\Omega} u_{it}\int_{-\infty}^{t} g_t(t-s)(u_i(t)-u_i(s))dsdx \\
&\quad \leq (n-2+\mid n-2\mid\varepsilon)\parallel u_{it}\parallel^2 \\
&\qquad + \frac{\mid n-2\mid g(0)\lambda_0^2}{4\varepsilon}\int_{-\infty}^{t}\mid g_t(t-s)\mid\parallel u_i(t)-u_i(s)\parallel^2_{H^1_{\Gamma_1}(\Omega)} ds.
\end{aligned}
\tag{7.4.31}
$$

Similarly to (7.4.30), using (7.2.8), (7.2.9), (7.4.7), boundary condition (7.2.13) and Young's inequality, we deduce

$$
\begin{aligned}
&\int_\Omega u_i u_{itt} dx \\
&= \int_\Omega u_i \Big(\mu\Delta(u_i - g * u_i) + (\lambda+\mu)\frac{\partial}{\partial x_i}(\mathrm{div}(u - g*u)) - \alpha\frac{\partial \theta_t}{\partial x_i}\Big)dx \\
&= \int_{\Gamma_2} u_i \Big(\mu\frac{\partial}{\partial \nu}(u_i - g*u_i) + (\lambda+\mu)\nu_i \mathrm{div}(u - g*u)\Big)d\Gamma \\
&\quad - \int_\Omega \Big(\mu\nabla u_i \nabla(u_i - g*u_i) + (\lambda+\mu)\mathrm{div}u\,\mathrm{div}(u-g*u)\Big)dx \\
&\quad - \alpha\int_\Omega u_i\frac{\partial\theta_t}{\partial x_i}dx \\
&= -\int_{\Gamma_2} u_i\Big(am\cdot\nu(u_i - g*u_i) + m\cdot\nu u_{it}\Big)d\Gamma \\
&\quad - \int_\Omega \Big(\mu\nabla u_i \nabla(u_i - g*u_i) + (\lambda+\mu)\mathrm{div}u\,\mathrm{div}(u-g*u)\Big)dx \\
&\quad - \alpha\int_\Omega u_i\frac{\partial\theta_t}{\partial x_i}dx \\
&\le -\int_{\Gamma_2} am\cdot\nu u_i(u_i - g*u_i)d\,Gamma \\
&\quad + \frac{R_0\gamma^2}{4\varepsilon}\int_{\Gamma_2} m\cdot\nu \mid u_{it}\mid^2 d\Gamma + 2\varepsilon \parallel u_i(t)\parallel^2_{H^1_{\Gamma_1}(\Omega)} + \frac{\alpha^2\lambda_0^2}{4\varepsilon}\parallel \nabla\theta_t(t)\parallel^2 \\
&\quad - \int_\Omega \Big(\mu\nabla u_i \nabla(u_i - g*u_i) + (\lambda+\mu)\mathrm{div}u\,\mathrm{div}(u-g*u)\Big)dx. \qquad (7.4.32)
\end{aligned}
$$

Noting definitions (7.2.29)–(7.2.31) of C_2, C_3, C_4, we therefore derive from (7.2.18), (7.4.7), (7.4.20), (7.2.27)–(7.4.32) and Young's inequality that

$$
\begin{aligned}
\frac{dF}{dt} &\le -\int_\Omega \Big(\mu\nabla u_i \nabla(u_i - g*u_i) + (\lambda+\mu)\mathrm{div}u\,\mathrm{div}(u-g*u)\Big)dx \\
&\quad - \int_{\Gamma_2} am\cdot\nu u_i(u_i - g*u_i)d\Gamma + (\varepsilon(1+\mid n-2\mid) - 2)\parallel u_{it}\parallel^2 \\
&\quad + \varepsilon(3 + 2\mid n-2\mid)\parallel u_i(t)\parallel^2_{H^1_{\Gamma_1}(\Omega)} + C_2\parallel\nabla\theta_t(t)\parallel^2 + C_3\int_{\Gamma_2} m\cdot\nu\mid u_{it}\mid^2 d\Gamma \\
&\quad + C_4\int_{-\infty}^t \mid g_t(t-s)\mid\parallel u_i(t) - u_i(s)\parallel^2_{H^1_{\Gamma_1}(\Omega)} ds \\
&\quad + (\varepsilon + 2\mid n-2\mid\varepsilon)\int_{-\infty}^t g(t-s)\parallel u_i(s)\parallel^2_{H^1_{\Gamma_1}(\Omega)} ds \\
&\quad + \int_{\Gamma_2} m\cdot\nu\Big(\frac{2a^2R_0^2}{\mu} - (n-2)a\Big)\mid u_i - g*u_i\mid^2 d\Gamma \\
&\le -E(u,\theta,t) + \frac{\alpha}{2\beta}(\parallel\theta_t(t)\parallel^2 + \parallel\nabla\theta(t)\parallel^2) - \kappa\parallel u_i(t)\parallel^2_{H^1_{\Gamma_1}(\Omega)} \\
&\quad + \frac{d}{dt}\int_{-\infty}^t G(t-s)\parallel u_i(s)\parallel^2_{H^1_{\Gamma_1}(\Omega)} ds
\end{aligned}
$$

$$
\begin{aligned}
&+[\varepsilon(1+\mid n-2\mid)-\frac{1}{2}]\parallel u_{it}(t)\parallel^2+\varepsilon(3+2\mid n-2\mid)\parallel u_i(t)\parallel^2_{H^1_{\Gamma_1}(\Omega)}\\
&+C_2\parallel\nabla\theta_t(t)\parallel^2+C_3\int_{\Gamma_2}m\cdot\nu\mid u_{it}\mid^2 d\Gamma\\
&+C_4\int_{-\infty}^t\mid g_t(t-s)\mid\parallel u_i(t)-u_i(s)\parallel^2_{H^1_{\Gamma_1}(\Omega)}ds\\
&+(\varepsilon+2\mid n-2\mid\varepsilon)\frac{d}{dt}\int_{-\infty}^t G(t-s)\parallel u_i(s)\parallel^2_{H^1_{\Gamma_1}(\Omega)}ds\\
&+\varepsilon(3+2\mid n-2\mid\varepsilon)(1-\kappa)\parallel u_i(t)\parallel^2_{H^1_{\Gamma_1}(\Omega)}\\
&+\int_{\Gamma_2}m\cdot\nu\Big(\frac{2a^2R_0^2}{\mu}-(n-2)a\Big)\mid u_i-g*u_i\mid^2 d\Gamma\\
\le&-E(u,\theta,t)+\Big(C_2+\frac{\alpha\lambda_0^2}{2\beta}\Big)\parallel\nabla\theta_t(t)\parallel^2+\frac{\alpha}{2\beta}\parallel\nabla\theta(t)\parallel^2+C_3\int_{\Gamma_2}m\cdot\nu\mid u_{it}\mid^2 d\Gamma\\
&+C_4\int_{-\infty}^t\mid g_t(t-s)\mid\parallel u_i(t)-u_i(s)\parallel^2_{H^1_{\Gamma_1}(\Omega)}ds\\
&+\Big(1+(\varepsilon+2\mid n-2\mid\varepsilon)\Big)\frac{d}{dt}\int_{-\infty}^t G(t-s)\parallel u_i(s)\parallel^2_{H^1_{\Gamma_1}(\Omega)}ds\\
&+(\varepsilon(1+\mid n-2\mid)-\frac{1}{2})\parallel u_{it}(t)\parallel^2\quad(=f_1)\\
&+\Big(\varepsilon(3+2\mid n-2\mid)+(\varepsilon+2\mid n-2\mid\varepsilon)(1-\kappa)-\kappa\Big)\parallel u_i(t)\parallel^2_{H^1_{\Gamma_1}(\Omega)}\quad(=f_2)\\
&+\int_{\Gamma_2}m\cdot\nu\Big(\frac{2a^2R_0^2}{\mu}-(n-2)a\Big)\mid u_i-g*u_i\mid^2 d\Gamma\quad(=f_3).
\end{aligned}
\tag{7.4.33}
$$

If $n\ge 3$, then, by (7.2.22), we deduce that $f_3\le 0$. In addition, by definition (7.2.34) of ε, we have $f_1\le 0$ and $f_2\le 0$. Hence, noting definition (7.2.32) of C_5, (7.4.19) follows from (7.4.33).

If $n\le 2$, then we estimate f_3 as follows. Let $K(A)$ be given by (7.2.33). Using (7.2.6), (7.2.8) and Young's inequality, we deduce

$$
\begin{aligned}
f_3\le&\ 2K(A)\int_{\Gamma_2}m\cdot\nu(\mid u_i\mid^2+\mid g*u_i\mid^2)d\Gamma\\
\le&\ 2K(A)R_0\gamma^2\parallel u_i(t)\parallel^2_{H^1_{\Gamma_1}(\Omega)}+2K(A)\int_{\Gamma_2}m\cdot\nu\Big(\int_{-\infty}^t g(t-s)u_i(s)ds\Big)^2 d\Gamma\\
\le&\ 2K(A)R_0\gamma^2\parallel u_i(t)\parallel^2_{H^1_{\Gamma_1}(\Omega)}\\
&+2K(A)\int_{\Gamma_2}m\cdot\nu\int_{-\infty}^t g(t-s)ds\int_{-\infty}^t g(t-s)\mid u_i(s)\mid^2 dsd\Gamma\\
\le&\ 2K(A)R_0\gamma^2\parallel u_i(t)\parallel^2_{H^1_{\Gamma_1}(\Omega)}\\
&+2K(A)(1-\kappa)\int_{\Gamma_2}m\cdot\nu\int_{-\infty}^t g(t-s)\mid u_i(s)\mid^2 dsd\Gamma\\
\le&\ 2K(A)R_0\gamma^2\parallel u_i(t)\parallel^2_{H^1_{\Gamma_1}(\Omega)}
\end{aligned}
$$

$$+2K(A)(1-\kappa)R_0\gamma^2\int_{-\infty}^{t} g(t-s) \parallel u_i(s) \parallel^2_{H^1_{\Gamma_1}(\Omega)} ds$$
$$= 2K(A)R_0\gamma^2(1+(1-\kappa)^2) \parallel u_i(t) \parallel^2_{H^1_{\Gamma_1}(\Omega)}$$
$$+2K(A)(1-\kappa)R_0\gamma^2\frac{d}{dt}\int_{-\infty}^{t} G(t-s) \parallel u_i(s) \parallel^2_{H^1_{\Gamma_1}(\Omega)} ds. \tag{7.4.34}$$

Set

$$C_6 = \varepsilon(3+2\mid n-2\mid) + (\varepsilon+2\mid n-2\mid\varepsilon)(1-\kappa) - \kappa + 2K(A)R_0\gamma^2(1+(1-\kappa)^2). \tag{7.4.35}$$

It therefore follows from (7.2.32), (7.4.33) and (7.4.34) that

$$\frac{dF}{dt}(u,\theta,t)$$
$$\leq -E(u,\theta,t) + (C_2 + \frac{\alpha\lambda_0^2}{2\beta}) \parallel \nabla\theta_t(t) \parallel^2 + \frac{\alpha}{2\beta} \parallel \nabla\theta(t) \parallel^2 + C_3\int_{\Gamma_2} m\cdot\nu \mid u_t\mid^2 d\Gamma$$
$$+ C_4\int_{-\infty}^{t} \mid g_t(t-s) \mid\parallel u_i(t) - u_i(s) \parallel^2_{(H^1_{\Gamma_1}(\Omega))^n} ds$$
$$+ C_5\frac{d}{dt}\int_{-\infty}^{t} G(t-s) \parallel u_i(s) \parallel^2_{(H^1_{\Gamma_1}(\Omega))^n} ds$$
$$+ \left(\varepsilon(1+\mid n-2\mid) - \frac{1}{2}\right) \parallel u_{it}(t) \parallel^2$$
$$+ C_6 \parallel u_i(t) \parallel^2_{(H^1_{\Gamma_1}(\Omega))^n}, \tag{7.4.36}$$

which implies (7.4.19) in view of definition (7.2.34) of ε. □

Lemma 7.4.4 *Let Γ_1 and Γ_2 be given by (7.1.15)–(7.1.16), respectively, satisfying (7.1.22)–(7.1.24). Suppose that the relaxation function g satisfies $(H_1) - (H_3)$. Assume that the function $a(x)$ satisfies (7.1.21) and (7.1.22). Let (u,θ) be the solution of problem (7.2.9)–(7.2.15). If*

$$0 < \delta \leq \min\left\{\frac{1}{C_3}, \frac{1}{C_4}, \frac{4\alpha(\lambda+\mu)\kappa}{2(\lambda+\mu)\kappa(2\beta C_2 + \alpha\lambda_0^2 + \alpha\lambda_1^2) + \alpha^2\lambda_1^2\beta}\right\}, \tag{7.4.37}$$

then we have

$$\tau V(u,\theta,t) \leq -V'(u,\theta,t) + \delta C_5\frac{d}{dt}\int_{-\infty}^{t} G(t-s) \parallel u(s) \parallel^2_{H^1_{\Gamma_1}(\Omega)^n} ds, \tag{7.4.38}$$

where

$$\tau = \frac{\delta}{4(1+\delta C_1)}, \tag{7.4.39}$$

and constants C_1, C_2, C_3, C_4, C_5 are given by (7.2.28)–(7.2.32), respectively.

Proof Multiplying (7.2.10) by θ in L^2, using Young's and Poincaré's inequalities, we conclude

$$\begin{aligned}\frac{dF_1(t)}{dt} &= \frac{\alpha}{2\beta} \parallel \theta_t(t) \parallel^2 + \frac{\alpha}{2}\int_\Omega \mathrm{div}u \cdot \theta_t dx - \frac{\alpha}{2\beta} \parallel \nabla\theta(t) \parallel^2 \\ &\le \frac{\lambda_1^2\alpha}{2\beta} \parallel \nabla\theta_t(t) \parallel^2 + \frac{(\lambda+\mu)\kappa}{4} \parallel \mathrm{div}u(t) \parallel^2 \\ &\quad + \frac{\lambda_1^2\alpha^2}{4(\lambda+\mu)\kappa} \parallel \nabla\theta_t(t) \parallel^2 - \frac{\alpha}{2\beta} \parallel \nabla\theta(t) \parallel^2 \end{aligned} \tag{7.4.40}$$

where λ_1 is the best constant in Poincaré's inequality. It follows from (7.2.19), (7.4.19), (7.4.40) and (H_2) that

$$\begin{aligned} V'(u,\theta,t) &= E'(u,\theta,t) + \delta\Big(F'(u,\theta,t) + F_1'(u,\theta,t)\Big) \\ &\le -\delta E(u,\theta,t) + \left(\delta(C_2 + \frac{\alpha\lambda_0^2}{2\beta} + \frac{\lambda_1^2\alpha}{2\beta} + \frac{\lambda_1^2\alpha^2}{4(\lambda+\mu)\kappa}) - \frac{\alpha}{\beta}\right) \parallel \nabla\theta_t(t) \parallel^2 \\ &\quad + \frac{(\lambda+\mu)\delta\kappa}{4} \parallel \mathrm{div}u(t) \parallel^2 + (\delta C_3 - 1)\int_{\Gamma_2} m\cdot\nu \mid u_{it} \mid^2 d\Gamma \\ &\quad + (\delta C_4 - 1)\int_{-\infty}^t \mid g_t(t-s) \mid \parallel u(t) - u(s) \parallel^2_{(H^1_{\Gamma_1}(\Omega))^n} ds \\ &\quad + \delta C_5 \frac{d}{dt}\int_{-\infty}^t G(t-s) \parallel u(s) \parallel^2_{(H^1_{\Gamma_1}(\Omega))^n} ds. \end{aligned} \tag{7.4.41}$$

Noting that (7.2.2) and (7.2.18), $\frac{(\lambda+\mu)\kappa}{4} \parallel \mathrm{div}u(t) \parallel^2$ is a part of $E(u,\theta,t)$, so we can choose $\delta' = \frac{\delta}{4}$ such that

$$-\delta E(u,\theta,t) + \frac{(\lambda+\mu)\delta\kappa}{4} \parallel \mathrm{div}u(t) \parallel^2 \le -\delta' E(u,\theta,t)$$

which, combined with (7.4.41), gives

$$\begin{aligned} V'(u,\theta,t) &= E'(u,\theta,t) + \delta\Big(F'(u,\theta,t) + F_1'(u,\theta,t)\Big) \\ &\le -\delta' E(u,\theta,t) + \left(\delta(C_2 + \frac{\alpha\lambda_0^2}{2\beta} + \frac{\lambda_1^2\alpha}{2\beta} + \frac{\lambda_1^2\alpha^2}{4(\lambda+\mu)\kappa}) - \frac{\alpha}{\beta}\right) \parallel \nabla\theta_t(t) \parallel^2 \\ &\quad + (\delta C_3 - 1)\int_{\Gamma_2} m\cdot\nu \mid u_t(t) \mid^2 d\Gamma \\ &\quad + (\delta C_4 - 1)\int_{-\infty}^t \mid g_t(t-s) \mid \parallel u(t) - u(s) \parallel^2_{(H^1_{\Gamma_1}(\Omega))^n} ds \\ &\quad + \delta C_5 \frac{d}{dt}\int_{-\infty}^t G(t-s) \parallel u(s) \parallel^2_{(H^1_{\Gamma_1}(\Omega))^n} ds. \end{aligned} \tag{7.4.42}$$

Thus (7.4.42) combining (7.4.5) and (7.4.37) implies (7.4.38). □

It is well known that, for a dynamical system $\{S(t)\}_{t\geq 0}$ on metric space X, a Lyapunov functional usually has the following property

$$V'(S(t)x) \leq -W(S(t)x) \leq 0, \quad for \ \ all \ x \in X, \tag{7.4.43}$$

where W is a given function. Because in (7.4.38) there is an additional term

$$\delta C_5 \frac{d}{dt} \int_{-\infty}^{t} G(t-s) \parallel u(s) \parallel^2_{(H^1_{\Gamma_1}(\Omega))^n} ds,$$

the functional V does not satisfies (7.4.43). Thus, it is referred to as a generalized Lyapunov functional.

Proof of Theorem 7.2.1 Multiplying both sides of (7.4.38) by $e^{\tau t}$ and integrating over $[0, t]$, we obtain

$$\begin{aligned}
V(u,\theta,t)e^{\tau t} - V(u,\theta,0) &\leq \delta C_5 \int_0^t e^{\tau r} dr \frac{d}{dr} \int_{-\infty}^{r} G(r-s) \parallel u(s) \parallel^2_{(H^1_{\Gamma_1}(\Omega))^n} ds \\
&= \delta C_5 e^{\tau t} \int_{-\infty}^{t} G(t-s) \parallel u(s) \parallel^2_{(H^1_{\Gamma_1}(\Omega))^n} ds \\
&\quad - \delta C_5 \int_{-\infty}^{0} G(t-s) \parallel u(s) \parallel^2_{(H^1_{\Gamma_1}(\Omega))^n} ds \\
&\quad - \delta C_5 \tau \int_0^t e^{\tau r} dr \int_{-\infty}^{r} G(r-s) \parallel u(s) \parallel^2_{(H^1_{\Gamma_1}(\Omega))^n} ds \\
&\leq \delta C_5 K \int_{-\infty}^{0} g(-s) \parallel u(s) \parallel^2_{(H^1_{\Gamma_1}(\Omega))^n} ds + \delta C_5 \tau K \int_0^t e^{\tau r} dr \\
&\quad \times \int_{-\infty}^{r} g(r-s) \parallel u(s) \parallel^2_{(H^1_{\Gamma_1}(\Omega))^n} ds \\
&\leq \delta C_5 K \int_{-\infty}^{0} g(-s) \parallel u(s) - u(0) + u(0) \parallel^2_{(H^1_{\Gamma_1}(\Omega))^n} ds \\
&\quad + \delta C_5 \tau K \int_0^t e^{\tau r} dr \int_{-\infty}^{r} g(r-s) \parallel u(s) - u(r) + u(r) \parallel^2_{(H^1_{\Gamma_1}(\Omega))^n} ds \\
&\leq 2\delta C_5 K \int_{-\infty}^{0} g(-s) \parallel u(0) \parallel^2_{(H^1_{\Gamma_1}(\Omega))^n} ds \\
&\quad + 2\delta C_5 K \int_{-\infty}^{0} g(-s) \parallel u(s) - u(0) \parallel^2_{(H^1_{\Gamma_1}(\Omega))^n} ds \\
&\quad + 2\delta C_5 \tau K \int_0^t e^{\tau r} dr \int_{-\infty}^{r} g(r-s) \parallel u(r) \parallel^2_{(H^1_{\Gamma_1}(\Omega))^n} ds \\
&\quad + 2\delta C_5 \tau K \int_0^t e^{\tau r} dr \int_{-\infty}^{r} g(r-s) \parallel u(s) - u(r) \parallel^2_{(H^1_{\Gamma_1}(\Omega))^n} ds \\
&\leq 2\delta C_5 K \frac{1-\kappa}{\kappa} E(u,\theta,0) + 2\delta C_5 K E(u,\theta,0) \\
&\quad + 2\delta C_5 K \frac{1-\kappa}{\kappa} (e^{\tau t} - 1) E(u,\theta,0) + 2\delta C_5 K (e^{\tau t} - 1) E(u,\theta,0) \\
&= \frac{2}{\kappa} \delta C_5 K e^{\tau t} E(u,\theta,0).
\end{aligned} \tag{7.4.44}$$

It therefore follows from (7.4.5) that

$$(1-\delta C_1)E(u,\theta,t) \le e^{-\tau t}(1+\delta C_1)E(u,\theta,0) + \frac{2}{\kappa}\delta C_5 K E(u,\theta,0)$$

and then

$$E(u,\theta,t) \le \rho E(u,\theta,0) \tag{7.4.45}$$

where

$$\rho(t) = \frac{e^{-\tau t}(1+\delta C_1)\kappa + 2\delta C_5 K}{\kappa(1-\delta C_1)}. \tag{7.4.46}$$

If δ is small enough and t is large enough so that

$$\frac{2\delta C_5 K}{\kappa(1-\delta C_1)} \le \frac{1}{2} \tag{7.4.47}$$

and

$$\frac{e^{-\tau t}(1+\delta C_1)}{(1-\delta C_1)} < \frac{1}{2}, \tag{7.4.48}$$

then $\rho < 1$. This holds if $t = T$ and δ is given by (7.2.26) and (7.2.27), respectively. Let $S(t)$ be the semigroup generated by system (7.2.9)–(7.2.15). Then (7.4.45) implies that

$$\| S(t) \| \le \rho. \tag{7.4.49}$$

Let $t = nT + s,\ 0 \le s < T$. Then

$$\| S(t) \| \le \| S(s) \| \| S(nT) \| \le \| S(T)^n \| \le (\rho(T))^n,$$

which implies that

$$\| S(t) \| \le (\rho(T))^{-1} e^{-\omega t}$$

where

$$\omega = -T^{-1}\ln\rho(T) > 0. \tag{7.4.50}$$

Hence, the proof of Theorem 7.2.1 is complete. □

7.5 Bibliographic Comments

There are some interesting results for the thermoelastic system of type III, Zhang and Zuazua [285], Quintanilla and Racke [233] independently studied the decay of energy for the problem of the linear thermoelastic system of type III by using the classical energy method and the spectral method, and obtained the exponential stability in one space dimension, and in two or three space dimensions for radially symmetric situations while the energy decays polynomially for most domains in two space dimensions; Quintanilla [231] proved how the solution of thermoelasticity of type III converges to the solution of the usual thermoelasticity and to the solution of thermoelasticity without energy dissipation, consequently he could also obtain an estimate of the approximation when he compared the classical thermoelasticity with thermoelasticity without energy dissipation. Quintanilla [229] investigated the structural stability in the class of thermoelasticity of type III. Recently, Messaoudi and Said-Houari [177] proved the energy decay in a transmission problem in thermoelasticity of type III, and also considered a Timoshenko-type system of thermoelasticity of type III [175] to prove an exponential decay result. The Cauchy problem of the linear thermoelastic system of type III was also studied by Reissig and Wang [243], Zhang and Zuazua [285], Yang and Wang [285]; Lazzari and Nibbi [140] studied the asymptotic behavior of the solutions of a three-dimensional thermoelastic systems of types II and III.

However, in the higher dimensional space case, it is well known that the energy, in general, does not tend to zero as $t \to +\infty$. Indeed, Lebeau and Zuazua [143] gave a sufficient and necessary condition ensuring that the energy tends to zero exponentially as $t \to +\infty$ in a bounded multi-dimensional smooth domain Ω. This condition is written in terms of the dynamics of the rays of geometric optics. As a consequence of the result of [197], it follows that when Ω is a bounded smooth convex open set, the energy does not decay exponentially to zero. This is because the total energy is not dissipative completely in the form of thermal energy. Therefore, in order to ensure the exponential stabilization in such case, a boundary velocity feedback was introduced in Liu [148] to increase the loss of energy for the model of thermoviscoelasticity.

If the relaxation function $g(t)$ satisfies conditions (H_1), (H_2) and (H_3), Navarro [196] considered the following model

$$\begin{cases} u_{tt} - \mu\Delta u - (\lambda+\mu)\nabla \mathrm{div} u \\ \quad +\mu g * \Delta u + (\lambda+\mu) g * \nabla \mathrm{div} u + \alpha\nabla\vartheta = 0, & (x,t) \in \Omega \times (0,+\infty), \\ \vartheta_t - \Delta\vartheta + \beta \mathrm{div} u_t = 0, & (x,t) \in \Omega \times (0,+\infty), \end{cases} \tag{7.5.1}$$

and proved the asymptotic stability for system (7.5.1), that is,

$$\lim_{t\to+\infty} \| u(t) \|_{(H_0^1(\Omega))^n} = \lim_{t\to+\infty} \| u'(t) \|_{(L^2(\Omega))^n} = \lim_{t\to+\infty} \| \vartheta(t) \|_{L^1(\Omega)} = 0.$$

Later on, Liu [148] proved the exponential stability with boundary feedback, and obtained the explicit decay rate. In the case of higher space dimension, the problem is very complicated. In order to see such complexity, we look at some special cases: thermoelastic system and viscoelastic system.

When $g = 0$, system (7.1.1)–(7.1.5) reduces to the thermoelastic system of type III without memory effects and boundary feedback, which was studied by Zhang an Zuazua [285]:

$$\begin{cases} u_{tt} - \mu\Delta u - (\lambda+\mu)\nabla \mathrm{div} u + \alpha\nabla\vartheta = 0, \quad (x,t)\in\Omega\times(0,+\infty), & (7.5.2)\\ \vartheta_{tt} - \Delta\vartheta - \Delta\vartheta_t + \beta \mathrm{div} u_{tt} = 0, \quad (x,t)\in\Omega\times(0,+\infty), & (7.5.3)\\ u = 0, \quad \vartheta = 0, \quad (x,t)\in\Gamma\times(0,+\infty), & (7.5.4)\\ u(x,0) = u_0(x),\ u_t(x,0) = u_1(x),\ \vartheta(x,0) = \vartheta_0(x),\ x\in\Omega. & (7.5.5) \end{cases}$$

This model comes from Green and Naghdi's theory [89, 90] in which the authors re-examined the classical Fourier's law in thermoelasticity, and introduced the so-called models of thermoelasticity of types II and III, whose constitutive assumptions on the heat flux vector are different from the Fourier's law, based on which the thermoelastic model of type I was proposed.

When $\alpha = \beta = 0$, system (7.1.1)–(7.1.5) is decoupled into the following viscoelastic system

$$\begin{cases} u_{tt} - \mu\Delta u - (\lambda+\mu)\nabla \mathrm{div} u & \\ \quad +\mu g * \Delta u + (\lambda+\mu) g * \nabla \mathrm{div} u = 0, \quad (x,t)\in\Omega\times(0,+\infty), & (7.5.6)\\ u = 0, \quad (x,t)\in\Gamma\times(0,+\infty), & (7.5.7)\\ u(x,0) = u_0(x),\ u_t(x,0) = u_1(x), \quad (x,t)\in\Omega, & (7.5.8)\\ u(x,0) - u(x,-s) = w_0(x), \quad (x,t)\in\Omega\times(0,+\infty), & (7.5.9) \end{cases}$$

and

$$\vartheta_{tt} - \Delta\vartheta_t - \Delta\vartheta = 0.$$

The energy $E(u,\vartheta,t)$ of problem (7.5.6–7.5.9) is defined by

$$\begin{aligned} E(u,\vartheta,t) = {} & \frac{1}{2}\Big[\| u_t(t) \|^2 + \mu\kappa \| \nabla u(t) \|^2 + (\lambda+\mu)\kappa \| \mathrm{div} u(t) \|^2 \Big] \\ & + \frac{\mu}{2}\int_{-\infty}^{t} g(t-s) \| \nabla u(t) - \nabla u(s) \|^2 \, ds \\ & + \frac{\lambda+\mu}{2}\int_{-\infty}^{t} g(t-s) \| \mathrm{div} u(t) - \mathrm{div} u(s) \|^2 \, ds. \end{aligned}$$

There are many results (see, e.g., [36, 57, 68, 121, 146, 154, 256, 277]) on exponential stability of energy and asymptotic stability of solutions to viscoelastic and thermoviscoelastic systems under different assumptions.

It can be easily seen from (7.1.13) that the asymptotic behavior of the energy $E(u,\theta,t)$ for (7.1.11) depends significantly on the relaxation function $g(t)$ and the history $w^0(x,s)$ as well. Subsequently, extensive attention was paid to the problem of obtaining an explicit decay rate. In this aspect, Day [53] first obtained a decay rate of t^{-1} in the case of one space dimension by introducing a feedback at one end of an interval. Later on, in the case of two space dimension, Leugering [141] established an exponential decay rate by introducing a velocity feedback on a part of the boundary of domain. On the other hand, if $g(t)$ decays exponentially and the initial history w_0 is taken to be zero, Desch and Miller [57] proved that, in the case of one space dimension, the energy also decays to zero exponentially at a rate no better than $g(t)$ decays.

In view of the above, in order to obtain an explicit decay rate of energy of higher dimensional thermoelastic system (7.1.11), it may be indispensable to introduce a velocity feedback on a part of the boundary of a thermoelastic body. Thus, in this chapter, we introduce such a feedback to increase the loss of energy and establish the exponential stabilization. Similar boundary velocity feedbacks were extensively used for the wave equation [35, 129], elastodynamic systems [135], viscoelasticity [104, 105, 141, 142], and thermoviscoelasticity [148].

For the models of thermoviscoelasticity and thermoplasticity, there are many results, see, e.g., [36,116, 117, 119, 121, 126, 131, 132, 133, 134, 154, 206, 207, 251, 256, 277, 288].

Chapter 8
Global Existence for the Three-Dimensional Thermoelastic Equations of Type II

8.1 Introduction

In this chapter, we shall study the global existence for the following thermoelastic model of type II:

$$\begin{cases} \rho u_{tt} = \nabla \cdot [C\nabla u(x,t) - \alpha I\theta(x,t)] + f, & (x,t) \in \Omega \times (0,+\infty), \quad (8.1.1) \\ c\theta_t = -\nabla \cdot [q(x,t) + \beta u_t(x,t)] + r, & (x,t) \in \Omega \times (0,+\infty). \quad (8.1.2) \end{cases}$$

We adopt the results in this chapter from [211].

From the Green–Naghdi law, we have

$$q(x,t) = -k\nabla\tau(x,t), \qquad (x,t) \in \Omega \times (0,+\infty) \tag{8.1.3}$$

where q is the heat flux, and τ is a new variable, called thermal displacement, which satisfies $\tau_t = \theta$.

The body Ω is a bounded open set of $\mathbb{R}^3$ with regular boundary $\partial\Omega$. Function $u(x,t) = (u_1(x,t), u_2(x,t), u_3(x,t))$ represents displacement vector, and θ denotes the temperature relative to Θ_0, i.e., $\theta = \Theta - \Theta_0$, where Θ represents the absolute temperature. Functions f and r represent external force, ρ represents the mass density and is a positive constant, and C is a constant, fourth-order, symmetric tensor which is positively defined, i.e., there exist two positive constants k_1 and k_2 such that, for all symmetric second-order tensors B,

$$k_1|B|^2 \le CB \cdot B \le k_2|B|^2 \tag{8.1.4}$$

while $c,\ k$, and $\frac{\alpha}{\beta}$ are positive constants.

Assume that system (8.1.1)–(8.1.3) subject to the following dissipative boundary condition with memory

Y. Qin and Z. Ma, *Global Well-posedness and Asymptotic Behavior of the Solutions to Non-classical Thermo(visco)elastic Models*,
DOI 10.1007/978-981-10-1714-8_8

$$T(x,t)n(x,t) = -\gamma_0 v(x,t) - \int_0^{+\infty} \lambda(s) v^t(x,s), \quad x \in \partial\Omega \tag{8.1.5}$$

and the Neumann boundary condition for the heat flux, that is,

$$q(x,t) \cdot n(x) = 0, \quad x \in \partial\Omega, \tag{8.1.6}$$

where n is the unit outward normal vector, $v := u_t$ the velocity, $v^t(x,s) := v(x,t-s)$ the history of v, and T is the stress tensor, which follows the constitutive equation

$$T = CE - \alpha I\theta \tag{8.1.7}$$

where $E = \frac{1}{2}(\nabla u + \nabla u^T)$ is the strain tensor and I is the identity tensor.

For boundary condition (8.1.5) with memory terms, several authors have studied the dynamical problem in elasticity (see, e.g., [20, 29, 186]) and in electromagnetism ([197]), and in thermoelasticity for the Cattaneo–Maxwell and Gurtin–Pipkin models ([23, 44]) and Green–Naghdi model of type II ([91, 140]). In this chapter, boundary condition (8.1.5) guarantees the decay of total energy (mechanical and thermal), because there is no internal dissipation for Green–Naghdi type II.

We set the initial conditions

$$\begin{cases} u(x,0) = u_0(x), \\ v(x,0) = v_0(x), \\ \tau(x,0) = \tau_0(x), \\ \theta(x,0) = \theta_0(x). \end{cases} \tag{8.1.8}$$

For the memory kernel $\lambda : \mathbb{R}^+ \to \mathbb{R} \in L^1(\mathbb{R}^+) \bigcap H^2(\mathbb{R}^+)$ and, by assuming that $\partial\Omega$ is locally strong dissipative ([3]), we have

$$\gamma_0 \in \mathbb{R}^+, \quad \omega \int_0^{+\infty} \lambda'(s) sin(\omega s)\, ds < 0, \quad for \ \ all \ \omega \neq 0.$$

Furthermore, we assume

$$\lambda'(s) < 0, \quad \lambda''(s) \geq 0, \quad for \ \ all \ s \in \mathbb{R}^+. \tag{8.1.9}$$

Models of boundary conditions including a memory term which produces damping were proposed in [17] for the study of 1D wave propagation, in [204] for sound evolution in a compressible fluid and in [69] in the context of Maxwell equations.

The notation in this chapter is standard. We put $\|\cdot\| = \|\cdot\|_{L^2(\Omega)}$, and denote by $C^k(J, B)$, $k \in \mathbb{N}_0$, the space of k-times continuously differentiable functions from $J \subseteq \mathbb{R}$ into a Banach space B, and likewise by $L^p(J, B)$, $1 \leq p \leq +\infty$, the corresponding Lebesgue spaces. $C^\beta([0, T], B)$ denotes the Hölder space of B-valued continuous functions with exponent $\beta \in (0, 1)$ in variable t.

From now on, we shall drop the x variable whenever no ambiguity arises. In the sequel, we shall refer to problem (8.1.1)–(8.1.3), (8.1.5)–(8.1.6) and (8.1.8) as to problem P. We adopt the results in this chapter from [227].

8.2 Main Results

Using integration by parts for (8.1.5), we obtain

$$T(x,t)n(x,t) = -\gamma_0 v(x,t) - \int_0^{+\infty} \lambda'(s) w^t(x,s)\, ds, \quad x \in \partial\Omega \tag{8.2.1}$$

where $w^t(x,s) = u^t(x,s) - u(x,t)$ denotes the past history of u and is defined for all $s \in \mathbb{R}^+$.

In order to simplify the notation, we introduce the new variable

$$a^t(s) = -\int_0^{+\infty} \lambda'(\tau + s) w^t(\tau)\, d\tau,$$

so that the boundary condition (8.1.5) or (8.2.1) takes the form

$$\check{T}(t)n = T(t)n + \gamma_0 v(t) = a^t(0) \tag{8.2.2}$$

and introduce the boundary energy function

$$\psi_{\partial\Omega}(t) = -\frac{1}{2}\int_{\partial\Omega}\int_0^{+\infty} \frac{1}{\lambda'(s)} \frac{\partial a^t(s)}{\partial s} \cdot \frac{\partial a^t(s)}{\partial s}\, ds da \tag{8.2.3}$$

which satisfies

$$\begin{aligned}\frac{d}{dt}\psi_{\partial\Omega}(t) &= \frac{1}{2}\int_{\partial\Omega} \frac{1}{\lambda'(0)} \frac{\partial a^t(0)}{\partial s} \cdot \frac{\partial a^t(0)}{\partial s}\, da \\ &\quad - \int_{\partial\Omega} \check{T}(t)n \cdot v(t)\, da - \frac{1}{2}\int_{\partial\Omega}\int_0^{+\infty} \frac{\lambda''(s)}{(\lambda'(s))^2} \frac{\partial a^t(s)}{\partial s} \cdot \frac{\partial a^t(s)}{\partial s}\, ds da.\end{aligned} \tag{8.2.4}$$

Thus the energy of solutions of problem P is defined by

$$\begin{aligned}\psi &= \psi_\Omega + \psi_{\partial\Omega} \\ &= \frac{1}{2}\int_\Omega \left(\rho |v|^2 + C\nabla u \cdot \nabla u + \frac{c\alpha}{\beta}\theta^2 + \frac{k\alpha}{\beta}|\nabla\tau|^2 \right) dx + \psi_{\partial\Omega}.\end{aligned} \tag{8.2.5}$$

We further introduce some function spaces. Set

$$H^*(\Omega) = \left\{ \Phi \in L^2(\Omega);\ \nabla \cdot \Phi \in L^2(\Omega) \right\}. \tag{8.2.6}$$

Let the space $\dot{H}^1(\partial\Omega) := \dot{H}^1\left(\partial\Omega \times (0,+\infty), \frac{1}{\sqrt{-\lambda'(s)}}\, dsda\right)$ consist of functions $a^t(s)$ on $(0,+\infty)$ for which

$$\|a^t\|^2_{\dot{H}^1} = -\int_{\partial\Omega}\int_0^{+\infty} \frac{1}{\lambda'(s)} \frac{\partial a^t(s)}{\partial s} \cdot \frac{\partial a^t(s)}{\partial s}\, dsda. \tag{8.2.7}$$

Put

$$\mathcal{H} = (L^2(\Omega))^3 \times (L^2(\Omega))^3 \times L^2(\Omega) \times L^2(\Omega) \times \dot{H}^1(\partial\Omega) \tag{8.2.8}$$

with the energy norm

$$\begin{aligned} \|(v, \nabla u, \theta, \nabla\tau, a^t)\|^2_{\mathcal{H}} = &\int_\Omega \left(\rho|v|^2 + C\nabla u \cdot \nabla u + \frac{c\alpha}{\beta}\theta^2 + \frac{k\alpha}{\beta}|\nabla\tau|^2\right) dx \\ &- \int_{\partial\Omega}\int_0^{+\infty} \frac{1}{\lambda'(s)} \frac{\partial a^t(s)}{\partial s} \cdot \frac{\partial a^t(s)}{\partial s}\, dsda. \end{aligned} \tag{8.2.9}$$

In order to use the theory of semigroups, we set

$$v(x,t) = u_t(x,t), \quad \tau_t(x,t) = \theta(x,t), \quad a^t(s) = -\int_0^{+\infty} \lambda'(\tau+s) w^t(\tau)\, d\tau.$$

By a straightforward calculation, we obtain

$$\frac{\partial a^t(s)}{\partial s} = \frac{\partial a^t(s)}{\partial t} + \lambda(s) v(x,t). \tag{8.2.10}$$

Thus we can write problem P as follows:

$$\begin{cases} \rho v_t = \nabla \cdot [C\nabla u(x,t) - \alpha I\theta(x,t)] + f, & (x,t) \in \Omega \times (0,+\infty), \\ c\theta_t = -\nabla \cdot [q(x,t) + \beta v(x,t)] + r, & (x,t) \in \Omega \times (0,+\infty), \\ q(x,t) = -k\nabla\tau(x,t), & (x,t) \in \Omega \times (0,+\infty), \\ T(x,t)n(x,t) = -\gamma_0 v(x,t) - \int_0^{+\infty} \lambda'(s) w^t(x,s), & x \in \partial\Omega, \\ q(x,t) \cdot n(x) = 0, & x \in \partial\Omega, \\ w^t(x,s) = u^t(x,s) - u(x,t), & (x,t) \in \Omega \times (0,+\infty), \\ u(x,0) = u_0(x), v(x,0) = v_0(x), & x \in \Omega, \\ \tau(x,0) = \tau_0(x), \theta(x,0) = \theta_0(x), & x \in \Omega. \end{cases} \tag{8.2.11}$$

Now we define a linear unbounded operator A on $\mathcal{H}$ by

$$\begin{aligned} &A(v, \nabla u, \theta, \nabla\tau, a^t) \\ &= \left(\frac{1}{\rho}\nabla \cdot (C\nabla u - \alpha I\theta), \nabla v, -\frac{1}{c}\nabla \cdot (\beta - k\nabla\tau), \nabla\theta, \frac{\partial a^t(s)}{\partial s} - \lambda(s)v\right). \end{aligned} \tag{8.2.12}$$

Put

$$\Phi = (v, \nabla u, \theta, \nabla\tau, a^t), \quad K = (F, 0, G, 0, 0) \tag{8.2.13}$$

where $F = \frac{1}{\rho} f$, $G = \frac{1}{c} r$. Then system (8.2.11) can be formulated as an abstract first-order Cauchy problem as follows

$$\begin{cases} \dfrac{d\Phi}{dt} = A\Phi + K, \\ \Phi(0) = \Phi_0 \end{cases} \tag{8.2.14}$$

on the Hilbert space $\mathcal{H}$, where $\Phi_0 = (v_0, \nabla u_0, \theta_0, \nabla\tau_0, a^0)$. The domain of A is given by:

$$\begin{aligned} D(A) = \Big\{ &(v, \nabla u, \theta, \nabla\tau, a^t) \in \mathcal{H} : v \in (H^1(\Omega))^3, \quad \nabla u \in (H^*(\Omega))^3, \quad \theta \in H^1(\Omega), \\ &\nabla\tau \in H^*(\Omega), \quad \frac{\partial a^t(s)}{\partial s} - \lambda(s)v \in \dot{H}^1(\partial\Omega); \quad q(x,t)\cdot n(x) = 0, \quad x \in \partial\Omega; \\ &T(x,t)n(x,t) = -\gamma_0 v(x,t) - \int_0^{+\infty} \lambda'(s) w^t(x,s)\, ds, \quad x \in \partial\Omega \Big\}. \end{aligned} \tag{8.2.15}$$

Note that $H^*(\Omega)$ is given by (8.2.6). It is clear that $D(A)$ is dense in $\mathcal{H}$.

We now state our main theorems.

Theorem 8.2.1 *Suppose $F = F(\Phi)$ and $G = G(\Phi)$, $\Phi = (v, \nabla u, \theta, \nabla\tau, a^t)$, and $K = (F, 0, G, 0, 0)$ satisfies the global Lipschitz condition on $\mathcal{H}$, i.e., there is a positive constant L such that for all $\Phi_1, \Phi_2 \in \mathcal{H}$,*

$$\|K(\Phi_1) - K(\Phi_2)\|_{\mathcal{H}} \le L\|\Phi_1 - \Phi_2\|_{\mathcal{H}}. \tag{8.2.16}$$

Then for any $\Phi_0 = (v_0, \nabla u_0, \theta_0, \nabla\tau_0, a^0) \in \mathcal{H}$, there exists a global mild solution Φ to system (8.2.11) *such that $\Phi \in C([0, +\infty), \mathcal{H})$, i.e.,*

$$\begin{cases} v(t) \in C([0,+\infty), (L^2(\Omega))^3); & \nabla u(t) \in C([0,+\infty), (L^2(\Omega))^3); \\ \theta(t) \in C([0,+\infty), L^2(\Omega)); & \nabla\tau(t) \in C([0,+\infty), L^2(\Omega)); \\ a^t(t) \in C([0,+\infty), \dot{H}^1(\partial\Omega)). \end{cases}$$

Theorem 8.2.2 *Suppose $F = F(\Phi)$ and $G = G(\Phi)$, $\Phi = (v, \nabla u, \theta, \nabla\tau, a^t)$, and $K = (F, 0, G, 0, 0)$ is a nonlinear operator from $D(A)$ into $D(A)$, and satisfies the global Lipschitz condition on $D(A)$, i.e., there is a positive constant L such that for all $\Phi_1, \ \Phi_2 \in D(A)$,*

$$\|K(\Phi_1) - K(\Phi_2)\|_{D(A)} \le L\|\Phi_1 - \Phi_2\|_{D(A)}. \tag{8.2.17}$$

Then for any $\Phi_0 = (v_0, \nabla u_0, \theta_0, \nabla\tau_0, a^0) \in D(A)$, there exists a unique global classical solution $\Phi = (v, \nabla u, \theta, \nabla\tau, a^t) \in C^1([0,+\infty), \ \mathcal{H}) \cap C([0,+\infty), D(A))$ to system (8.2.11)*, i.e.,*

$$\begin{cases} v \in C^1([0,+\infty), (L^2(\Omega))^3) \cap C([0,+\infty), (H^1(\Omega))^3), \\ \nabla u \in C^1([0,+\infty), (L^2(\Omega))^3) \cap C([0,+\infty), (H^*(\Omega))^3), \\ \theta \in C^1([0,+\infty), L^2(\Omega)) \cap C([0,+\infty), H^1(\Omega)), \\ \nabla\tau \in C^1([0,+\infty), L^2(\Omega)) \cap C([0,+\infty), H^*(\Omega)), \\ a^t \in C^1([0,+\infty), \dot{H}^1(\partial\Omega)) \cap C([0,+\infty), \dot{H}^2(\partial\Omega)) \end{cases}$$

where $\dot{H}^2(\partial\Omega) = \dot{H}^2\left((\partial\Omega \times (0,+\infty), \frac{1}{\sqrt{-\lambda'(s)}}\, dsda\right)$ *with the norm*

$$\|a^t\|_{\dot{H}^2}^2 = -\int_{\partial\Omega}\int_0^{+\infty} \frac{1}{\lambda'(s)} \frac{\partial^2 a^t(s)}{\partial s^2} \cdot \frac{\partial^2 a^t(s)}{\partial s^2}\, dsda.$$

8.3 Proofs of Theorems 8.2.1–8.2.2

In this section, we shall complete proofs of Theorems 8.2.1–8.2.2 via the Lumer–Phillips theorem in terms of establishing a series of lemmas.

Lemma 8.3.1 *The operator A defined by* (8.2.12) *is dissipative and closed.*

Proof By a straightforward calculation, it follows from (8.2.2), (8.2.9), (8.2.12) and boundary conditions (8.1.5)–(8.1.6) that for any $(v, \nabla u, \theta, \nabla\tau, a^t) \in D(A)$,

$$\begin{aligned}
&(A(v, \nabla u, \theta, \nabla\tau, a^t),\ (v, \nabla u, \theta, \nabla\tau, a^t)) \\
&= \int_\Omega \Big[\nabla\cdot(C\nabla u(t) - \alpha I\theta(t))\cdot v(t) + C\nabla v(t)\cdot\nabla u(t)\Big]dx \\
&\quad + \frac{\alpha}{\beta}\int_\Omega \Big[\nabla\cdot(k\nabla\tau(t) - \beta v(t))\,\theta(t) + k\nabla\theta(t)\cdot\nabla\tau(t)\Big]dx \\
&\quad - \int_{\partial\Omega}\int_0^{+\infty} \frac{1}{\lambda'(s)}\frac{\partial}{\partial s}\left[\frac{\partial a^t(s)}{\partial s} - \lambda(s)v(t)\right]\cdot\frac{\partial a^t(s)}{\partial s}\,dsda \\
&= \int_{\partial\Omega}(C\nabla u - \alpha I\theta)\,v\cdot n\,da - \frac{\alpha}{\beta}\int_{\partial\Omega}(\beta v - k\nabla\tau)\theta\cdot n\,da \\
&\quad + \alpha\int_\Omega I\theta\cdot\nabla v\,dx + \frac{\alpha}{\beta}\int_\Omega(\beta v - k\nabla\tau)\cdot\nabla\theta\,dx + \frac{\alpha}{\beta}\int_\Omega k\nabla\theta\cdot\nabla\tau\,dx \\
&\quad - \int_{\partial\Omega}\int_0^{+\infty}\frac{1}{\lambda'(s)}\frac{\partial^2 a^t(s)}{\partial s^2}\cdot\frac{\partial a^t(s)}{\partial s}\,dsda + \int_{\partial\Omega}\int_0^{+\infty} v(t)\cdot\frac{\partial a^t(s)}{\partial s}\,dsda \\
&= \int_{\partial\Omega} T(t)n\cdot v(t)\,da - \int_{\partial\Omega} v(t)\cdot a^t(0)\,da - \int_{\partial\Omega}\int_0^{+\infty}\frac{1}{\lambda'(s)}\frac{\partial^2 a^t(s)}{\partial s^2}\cdot\frac{\partial a^t(s)}{\partial s}\,dsda \\
&= -\int_{\partial\Omega}\gamma_0|v(t)|^2\,da + \frac{1}{2}\int_{\partial\Omega}\frac{1}{\lambda'(0)}\frac{\partial a^t(0)}{\partial s}\cdot\frac{\partial a^t(0)}{\partial s}\,da \\
&\quad - \frac{1}{2}\int_{\partial\Omega}\int_0^{+\infty}\frac{\lambda''(s)}{(\lambda'(s))^2}\frac{\partial a^t(s)}{\partial s}\cdot\frac{\partial a^t(s)}{\partial s}\,dsda \le 0.
\end{aligned} \tag{8.3.1}$$

Thus A is dissipative.

To prove that A is closed, let $(v_n, \nabla u_n, \theta_n, \nabla \tau_n, a_n^t) \in D(A)$ be such that

$$(v_n, \nabla u_n, \theta_n, \nabla \tau_n, a_n^t) \to (v, \nabla u, \theta, \nabla \tau, a^t) \quad \text{in} \quad \mathcal{H}$$

and

$$A(v_n, \nabla u_n, \theta_n, \nabla \tau_n, a_n^t) \to (\varphi, z, \xi, \eta, \zeta) \quad \text{in} \quad \mathcal{H}.$$

Then we have

$$\begin{cases} v_n \to v & \text{in} \quad (L^2(\Omega))^3, & (8.3.2)\\ \nabla u_n \to \nabla u & \text{in} \quad (L^2(\Omega))^3, & (8.3.3)\\ \theta_n \to \theta & \text{in} \quad L^2(\Omega), & (8.3.4)\\ \nabla \tau_n \to \nabla \tau & \text{in} \quad L^2(\Omega), & (8.3.5)\\ a_n^t \to a^t & \text{in} \quad \dot{H}^1(\partial\Omega) & (8.3.6) \end{cases}$$

and

$$\begin{cases} \frac{1}{\rho}\nabla \cdot (C\nabla u_n - \alpha I \theta_n) \to \varphi & \text{in} \quad (L^2(\Omega))^3, & (8.3.7)\\ \nabla v_n \to z & \text{in} \quad (L^2(\Omega))^3, & (8.3.8)\\ -\frac{1}{c}\nabla \cdot (\beta v_n - k\nabla \tau_n) \to \xi & \text{in} \quad L^2(\Omega), & (8.3.9)\\ \nabla \theta_n \to \eta & \text{in} \quad L^2(\Omega), & (8.3.10)\\ \frac{\partial a_n^t}{\partial s} - \lambda(s) v_n \to \zeta & \text{in} \quad \dot{H}^1(\partial\Omega). & (8.3.11) \end{cases}$$

By Eqs. (8.3.2) and (8.3.8), we deduce

$$v_n \to v \quad \text{in} \quad (H^1(\Omega))^3 \tag{8.3.12}$$

and

$$z = \nabla v, \quad v \in (H^1(\Omega))^3. \tag{8.3.13}$$

Similarly, by (8.3.4) and (8.3.10), we deduce

$$\theta_n \to \theta \quad \text{in} \quad H^1(\Omega) \tag{8.3.14}$$

and

$$\eta = \nabla \theta, \quad \theta \in H^1(\Omega). \tag{8.3.15}$$

By (8.3.7) and (8.3.14), we deduce

$$\nabla \cdot (C\nabla u_n) \to \rho\varphi + \alpha \nabla \cdot I\theta \quad \text{in} \quad (L^2(\Omega))^3. \tag{8.3.16}$$

Thus, it follows from (8.3.3) that

$$\nabla u_n \to \nabla u \quad \text{in} \quad (H^*(\Omega))^3, \tag{8.3.17}$$

and

$$\varphi = \frac{1}{\rho}\nabla \cdot (C\nabla u - \alpha I\theta), \quad \nabla u \in (H^*(\Omega))^3. \tag{8.3.18}$$

By (8.3.9) and (8.3.12), we deduce

$$\frac{1}{c}\nabla \cdot (k\nabla \tau_n) \to \xi + \frac{\beta}{c}\nabla \cdot v \quad \text{in} \quad L^2(\Omega), \tag{8.3.19}$$

and consequently, it follows from (8.3.5) that

$$\nabla \tau_n \to \nabla \tau \quad \text{in} \quad H^*(\Omega), \tag{8.3.20}$$

and

$$\xi = -\frac{1}{c}\nabla \cdot (C\nabla u - \alpha I\theta), \quad \nabla \tau \in H^*(\Omega). \tag{8.3.21}$$

In addition, it follows from (8.3.6), (8.3.11) and (8.3.12) that

$$\frac{\partial a_n^t}{\partial s} \to \frac{\partial a^t}{\partial s} \quad \text{in} \quad \dot{H}^1(\partial\Omega), \tag{8.3.22}$$

and

$$\zeta = \frac{\partial a^t}{\partial s} - \lambda(s)v \in \dot{H}^1(\partial\Omega). \tag{8.3.23}$$

Moreover, it is easy to deduce

$$\begin{cases} T(x,t)n(x,t) = -\gamma_0 v(x,t) - \displaystyle\int_0^{+\infty} \lambda'(s)w^t(x,s), \quad x \in \partial\Omega, \\ q(x,t)\cdot n(x) = 0, \qquad x \in \partial\Omega. \end{cases}$$

Thus by (8.3.13), (8.3.15), (8.3.18), (8.3.21) and (8.3.23), we deduce

$$A(v, \nabla u, \theta, \nabla\tau, a^t) = (\varphi, z, \xi, \eta, \zeta), \quad (v, \nabla u, \theta, \nabla\tau, a^t) \in D(A).$$

Hence, A is closed. □

Lemma 8.3.2 *The adjoint operator A^* of A is also dissipative.*

Proof Let $\tilde{\Phi} = (\tilde{v}, \nabla\tilde{u}, \tilde{\theta}, \nabla\tilde{\tau}, \tilde{a}^t)$ be in $\mathcal{H}$ and consider the boundary conditions

$$(C\nabla\tilde{u}(t) - \alpha I\tilde{\theta}(t))n = \gamma_0\tilde{u}(t) + \tilde{a}^t(0), \quad \nabla\tilde{\tau}(t)\cdot n = 0 \quad x \in \partial\Omega. \tag{8.3.24}$$

Denoting by H the Heaviside function and introducing a function $j(\tilde{a}^t)$ such that

$$\frac{\partial}{\partial s} j(\tilde{a}^t)(s) = -\lambda'(s)\frac{\partial}{\partial s}\left(\frac{H(s)}{\lambda'(s)}\right)\frac{\partial \tilde{a}^t}{\partial s},$$

we claim that $A^*\tilde{\Phi}$ is equal to

$$\left(\frac{1}{\rho}\nabla\cdot(\alpha I\tilde{\theta} - C\nabla\tilde{u}), -\nabla\tilde{v}, \frac{1}{c}\nabla\cdot(\beta\tilde{v} - k\nabla\tilde{\tau}), -\nabla\tilde{\theta}, -\frac{\partial \tilde{a}^t(s)}{\partial s} + \lambda(s)\tilde{v} + j(\tilde{a}^t)(s)\right)$$

and the domain of A^* is

$$\begin{aligned} D(A^*) = \Big\{&(\tilde{v}, \nabla\tilde{u}, \tilde{\theta}, \nabla\tilde{\tau}, \tilde{a}^t) \in \mathcal{H} : \tilde{v} \in (H^1(\Omega))^3;\quad \nabla\tilde{u} \in (H^*(\Omega))^3;\quad \tilde{\theta} \in H^1(\Omega);\\ &\nabla\tilde{\tau} \in H^*(\Omega);\quad -\frac{\partial \tilde{a}^t(s)}{\partial s} + \lambda(s)\tilde{v} + j(\tilde{a}^t)(s) \in \dot{H}^1(\Omega);\quad \nabla\tilde{\tau}(t)\cdot n = 0 \quad x \in \partial\Omega;\\ &(C\nabla\tilde{u}(t) - \alpha I\tilde{\theta}(t))n = \gamma_0\tilde{u}(t) + \tilde{a}^t(0), \quad x \in \partial\Omega\Big\}. \end{aligned} \tag{8.3.25}$$

By a straightforward calculation, we can obtain for any $\Phi \in D(A)$ and $\tilde{\Phi} \in D(A^*)$,

$$\begin{aligned} (A\Phi, \tilde{\Phi}) =& \int_\Omega \Big[\nabla\cdot(C\nabla u(t) - \alpha I\theta(t))\cdot\tilde{v}(t) + C\nabla v\cdot\nabla\tilde{u}(t)\Big]dx\\ &-\frac{\alpha}{\beta}\int_\Omega \Big[\nabla\cdot(\beta v(t) - k\nabla\tau(t))\,\tilde{\theta}(t) + k\nabla\theta(t)\cdot\nabla\tilde{\tau}(t)\Big]\,dx\\ &-\int_{\partial\Omega}\int_0^{+\infty}\frac{1}{\lambda'(s)}\frac{\partial}{\partial s}\left(\frac{\partial a^t(s)}{\partial s} - \lambda(s)v(t)\right)\cdot\frac{\partial\tilde{a}^t(s)}{\partial s}\,dsda\\ =& -\int_\Omega\Big[v(t)\cdot\nabla\cdot\Big(C\nabla\tilde{u}(t) - \alpha I\tilde{\theta}(t)\Big) + C\nabla u(t)\cdot\nabla\tilde{v}(t)\Big]dx\\ &+\frac{\alpha}{\beta}\int_\Omega\Big[\nabla\cdot(\beta\tilde{v}(t) - k\nabla\tilde{\tau}(t))\,\theta(t) + k\nabla\tilde{\theta}(t)\cdot\nabla\tau(t)\Big]\,dx\\ &+\int_{\partial\Omega} v(t)\cdot\Big[\Big(C\nabla\tilde{u}(t) - \alpha I\tilde{\theta}(t)\Big)n - \gamma_0\tilde{v}(t) - \tilde{a}^t(0)\Big]\,da\\ &+\int_{\partial\Omega}\int_0^{+\infty}\frac{1}{\lambda'(s)}\frac{\partial a^t(s)}{\partial s}\cdot\frac{\partial}{\partial s}\left(\frac{\partial\tilde{a}^t(s)}{\partial s} - \lambda(s)\tilde{v}(t)\right)\,dsda\\ &-\frac{k\alpha}{\beta}\int_{\partial\Omega}\theta(t)\nabla\tilde{\tau}(t)\cdot n\,da + \int_{\partial\Omega}\frac{1}{\lambda'(0)}\frac{\partial a^t(0)}{\partial s}\cdot\frac{\partial\tilde{a}^t(0)}{\partial s}\,da\\ &+\int_{\partial\Omega}\int_0^{+\infty}\frac{\partial}{\partial s}\left(\frac{1}{\lambda'(s)}\right)\frac{\partial a^t(s)}{\partial s}\cdot\frac{\partial\tilde{a}^t(s)}{\partial s}\,dsda. \end{aligned}$$

Note that $\tilde{\Phi} \in D(A^*)$, so $\tilde{\Phi} \in D(A^*)$ satisfies boundary conditions (8.3.24), and we have

$$(A\Phi, \tilde{\Phi}) = (\Phi, -A\tilde{\Phi}) + \int_{\partial\Omega}\int_{-\infty}^{+\infty} \frac{\partial}{\partial s}\left(\frac{H(s)}{\lambda'(s)}\right) \frac{\partial a^t(s)}{\partial s} \cdot \frac{\partial \tilde{a}^t(s)}{\partial s}\, dsda.$$

Thus

$$(A^*\tilde{\Phi}, \tilde{\Phi}) = -\int_{\partial\Omega} \gamma_0 |\tilde{v}(t)|^2\, da - \frac{1}{2}\int_{\partial\Omega}\int_0^{+\infty} \frac{\lambda''(s)}{(\lambda'(s))^2} \frac{\partial \tilde{a}^t(s)}{\partial s} \cdot \frac{\partial \tilde{a}^t(s)}{\partial s}\, dsda \le 0.$$

Hence, A^* is also dissipative. □

Lemma 8.3.3 *Let A be a densely defined linear operator on a Hilbert space $\mathcal{H}$, if the operator A and A^* (the adjoint of A) are dissipative, then A generates a C_0-semigroup of contractions on $\mathcal{H}$.*

Proof By virtue of Lumer–Phillips theorem (see, e.g., Pazy [201]), we need to prove

$$R(I - A) = \mathcal{H}, \tag{8.3.26}$$

that is, $I - A$ is surjective. If A is dissipative and closed, then $R(I - A) \subseteq \mathcal{H}$. Suppose that $R(I - A) \neq \mathcal{H}$, then there is a nontrivial element $x^* \in \mathcal{H}^*$ such that for all $x \in D(A)$,

$$(x^*, x - Ax) = 0. \tag{8.3.27}$$

Thus

$$x^* - A^*x^* = 0. \tag{8.3.28}$$

Since A^* is dissipative, we know $x^* = 0$, a contradiction. Hence the proof is complete. □

From Lemmas 8.3.1–8.3.3, we know that operator A defined by (8.2.12) generates a C_0-semigroup of contractions on $\mathcal{H}$.

If we choose $B = -A$, by virtue of Lemmas 8.3.1–8.3.3 and definition of maximal accretive operator, we know that operator B is a maximal accretive operator and generates a C_0-semigroup $S(t)$ of contractions on $\mathcal{H}$.

Then system (8.2.11) can be formulated as an abstract first-order Cauchy problem as follows:

$$\begin{cases} \dfrac{d\Phi}{dt} + B\Phi = K, \\ \Phi(0) = \Phi_0 \end{cases} \tag{8.3.29}$$

where $\Phi_0 = (v_0, \nabla u_0, \theta_0, \nabla\tau_0, a^0)$, B is a maximal accretive operator defined in a dense subset $D(B) = D(A)$ of a Hilbert space $\mathcal{H}$.

Similarly to Lemmas 5.3.2–5.3.5, we can easily show the following two lemmas (see, e.g., Zheng [287]).

Lemma 8.3.4 *Suppose that $K = K(t)$, and*

$$K(t) \in C^1([0, +\infty), \mathcal{H}), \quad \Phi_0 \in D(B).$$

Then problem (8.3.29) *admits a unique global classical solution u such that*

$$\Phi \in C^1([0, +\infty), H) \cap C([0, +\infty), D(B)) \tag{8.3.30}$$

which can be expressed as

$$\Phi(t) = S(t)\Phi_0 + \int_0^t S(t-\tau)K(\tau)d\tau. \tag{8.3.31}$$

Lemma 8.3.5 *Suppose that* $K = K(t)$, *and*

$$K(t) \in C([0, +\infty), D(B)), \quad \Phi_0 \in D(B). \tag{8.3.32}$$

Then problem (8.3.29) *admits a unique global classical solution.*

Proof of Theorem 8.2.1 The proof is similar to that of Theorem 5.2.3. □
Proof of Theorem 8.2.2 The proof is similar to that of Theorem 5.2.4. □

8.4 Bibliographic Comments

For the thermoelasticity of type II, or without energy dissipation, several results on existence, uniqueness, continuous dependence, spatial decay and wave propagation (see, e.g., [8, 9, 27, 34, 91, 135, 125, 140, 194, 226, 227, 230, 232]) have been obtained, among which we would like to mention the work by Qin and Muñoz Rivera [226] who studied the global existence and exponential stability of solutions to homogeneous thermoelastic equations of type II with thermal memory. Recently, Qin, Xu and Ma [227] obtained the global existence and exponential stability of solutions to non-homogeneous thermoelastic equations of type II with thermal memory. Lazzari and Nibbi [140] obtained the exponential decay of total energy to thermoelastic linear inhomogeneous system of type II (i.e., $f = f(x,t), r = r(x,t)$) with dissipative boundary condition (8.1.5), they used the Green–Naghdi theory of thermomechanics of continua to derive a nonlinear theory of thermoelasticity with diffusion of types II and III. This theory permits propagation of both thermal and diffusion waves at finite speeds. Liu and Quintanilla [152] studied the energy decay rate for a mixed type II and type III thermoelastic system by verifying the frequency domain conditions. The system consists of a wave equation and a heat equation of type III in one part of the domain, a wave equation and a heat equation of type II in another part of the domain, coupled in certain pattern. When the damping coefficient function satisfies certain conditions at the interface, a polynomial type decay rate is obtained. Later on, Aouadi, Lazzari and Nibbi [5] used the Green–Naghdi theory of thermomechanics of continua to derive a nonlinear theory of thermoelasticity with diffusion of types II and III with the help of the semigroup theory of linear operators, and established that the linear anisotropic problem is wellposed and we studied the asymptotic behavior of the solutions. Finally, they investigated the impossibility of the localization in time of solutions. We study in this chapter the global existence of solutions for the semilinear and nonlinear thermoelastic systems of type II (i.e., $f = f(v, \nabla u, \theta, \nabla \tau, a^t)$ and $f = f(v, \nabla u, \theta, \nabla \tau, a^t)$). To our knowledge, we first use the semigroup approach to study such a problem.

For the thermoelasticity of type III, which presents thermal dissipation, there are some interesting results (see, e.g., [233, 243, 282, 285]), for example, for the Cauchy problem of the linear thermoelastic system of type III, Zhang and Zuazua [285], and Quintanilla and Racke [233] independently studied the decay of energy by using the classical energy method and the spectral method, and obtained the exponential stability in one space dimension, and in two or three space dimensions for radially symmetric situations, while the energy decays polynomially for most domains in two space dimensions. Reissig and Wang [243] studied the $L^p - L^q$ decays estimates and propagation of singularities of solutions in one space dimension, and later on Yang and Wang [282] studied well-posedness and decay estimates in three space dimension. Lazzari and Nibbi [140] also studied the asymptotic behavior of the solution of a $3D$ thermoelastic system of type III with an absorbing boundary.

Chapter 9
Energy Decay for Thermoviscoelastic Systems

9.1 Introduction

In this chapter, we shall consider the following thermoviscoelastic problem

$$\begin{cases} u_{tt} - \Delta u + g * \Delta u + \nabla \vartheta = 0, & (x,t) \in \Omega \times (0, +\infty), \\ \vartheta_{tt} - \Delta \vartheta_t - \Delta \vartheta + \text{div} u_{tt} = 0, & (x,t) \in \Omega \times (0, +\infty), \\ \vartheta = 0, & (x,t) \in \partial\Omega \times (0, +\infty), \\ u = 0, & (x,t) \in \Gamma_0 \times (0, +\infty), \\ \frac{\partial u}{\partial \nu} - g * \frac{\partial u}{\partial \nu} + H(u_t) = 0, & (x,t) \in \Gamma_1 \times (0, +\infty) \end{cases} \tag{9.1.1}$$

with the initial data

$$\begin{aligned} &u(x,0) = u_0(x), \quad u_t(x,0) = u_1(x), \quad \vartheta(x,0) = \vartheta_0(x), \\ &\vartheta_t(x,0) = \vartheta_1(x), \quad x \in \Omega, \end{aligned} \tag{9.1.2}$$

where Ω is a bounded domain of $\mathbb{R}^n$ with a smooth boundary $\partial\Omega = \Gamma_0 \cup \Gamma_1$. Here following [173], Γ_0 and Γ_1 are closed and disjoint, with meas $(\Gamma_0) > 0$, ν is the unit outward normal to $\partial\Omega$, $u(x,t) = \left(u^1(x,t), \ldots, u^n(x,t)\right)$, and $\vartheta(x,t)$ represent displacement vector and temperature derivations, respectively, $g(t) = \left(g^1(t), \ldots, g^n(t)\right)$ is a specific vector function and $H = \left(h^1(u^1), \ldots, h^n(u^n)\right)$, $i = 1, 2, \ldots, n$, $h^i : \mathbb{R} \to \mathbb{R}$ are functions specified below, $g * \Delta u = \left(g^1 * \Delta u^1, \ldots, g^n * \Delta u^n\right)$, $g^i * \Delta u^i = \int_0^t g^i(t-\tau) u^i(\tau) d\tau$.

The basic conditions on the relaxation function $g^i(t)$ $(i = 1, \ldots, n)$ are standard (see, e.g., Dafermos [48–51] and [148])

Y. Qin and Z. Ma, *Global Well-posedness and Asymptotic Behavior of the Solutions to Non-classical Thermo(visco)elastic Models*,
DOI 10.1007/978-981-10-1714-8_9

$$\begin{cases} (H_1) & g^i \in C^1[0,+\infty) \cap L^1(0,+\infty), \quad i = 1,\ldots,n; \\ (H_2) & g^i(t) \geq 0, \quad (g^i)'(t) \leq 0, \quad t > 0, \quad i = 1,\ldots,n; \\ (H_3) & l_i = 1 - \displaystyle\int_0^{+\infty} g^i(t)\,dt > 0, \quad i = 1,\ldots,n. \end{cases}$$

Condition (H_3) simply states that the static modulus of elasticity is positive. This restriction is quite natural. In addition, conditions (H_1) and (H_2) imply

$$g^i(+\infty) = \lim_{t\to+\infty} g^i(t) = 0. \tag{9.1.3}$$

In the sequel, we denote by $\|\cdot\|$ the norm of $L^2(\Omega)$.
As in [285], we introduce the new variable

$$\theta(x,t) = \int_0^t \vartheta(x,s)ds + \chi(x) \tag{9.1.4}$$

with $\chi(x)$ to be defined below, instead of the temperature ϑ, of course, according to (9.1.4), we have

$$\theta_t = \vartheta. \tag{9.1.5}$$

Integrating the second equation in (9.1.1) with respect to t, we deduce

$$\theta_{tt} - \Delta\theta_t - \Delta\theta + \mathrm{div} u_t + \Delta\chi - \vartheta_1 + \Delta\vartheta_0 - \mathrm{div} u_1 = 0. \tag{9.1.6}$$

Therefore by choosing $\chi \in H_0^1(\Omega)$ such that

$$-\Delta\chi = \Delta\vartheta_0 - \mathrm{div} u_1 - \vartheta_1 \qquad in\ \Omega, \tag{9.1.7}$$

system (9.1.1)–(9.1.2) can be rewritten as the following system

$$\begin{cases} u_{tt} - \Delta u + g * \Delta u + \nabla\theta_t = 0, & (x,t) \in \Omega \times (0,+\infty), \\ \theta_{tt} - \Delta\theta_t - \Delta\theta + \mathrm{div} u_t = 0, & (x,t) \in \Omega \times (0,+\infty), \\ \theta = 0, & (x,t) \in \partial\Omega \times (0,+\infty), \\ u = 0, & (x,t) \in \Gamma_0 \times (0,+\infty), \\ \frac{\partial u}{\partial \nu} - g * \frac{\partial u}{\partial \nu} + H(u_t) = 0, & (x,t) \in \Gamma_1 \times (0,+\infty), \end{cases} \tag{9.1.8}$$

with the initial data

$$u(x,0) = u_0(x), \quad u_t(x,0) = u_1(x), \quad \theta(x,0) = \theta_0(x), \quad \theta_t(x,0) = \theta_1(x), \quad x \in \Omega. \tag{9.1.9}$$

Thus, in the sequel, we shall analyze system (9.1.8) instead of (9.1.1).

The energy $E(u,\theta)(t)$ of system (9.1.8) is defined by

$$E(u,\theta)(t) = \frac{1}{2}\int_\Omega \left(| u_t |^2 + \theta_t^2 + |\nabla\theta|^2 + |\nabla u|^2 - \left(\sum_{i=1}^n \int_0^t g^i(s)ds\, |\nabla u^i|^2 \right) \right) dx$$
$$+ \frac{1}{2}(g \circ \nabla u)(t), \tag{9.1.10}$$

where, for all $v \in L^2(\Omega)$,

$$(g \circ v)(t) = \sum_{i=1}^n \left(\int_\Omega \int_0^t g^i(t-s) \mid v^i(t) - v^i(s) \mid^2 dsdx \right). \tag{9.1.11}$$

Let

$$E_i(t) = \frac{1}{2}\int_\Omega \left(| u_t^i |^2 + \theta_t^2 + |\nabla\theta|^2 + \left(1 - \int_0^t g^i(s)ds |\nabla u^i(t)|^2 \right) \right) dx$$
$$+ \frac{1}{2}(g^i \circ \nabla u^i)(t), \quad i = 1, 2, \ldots, n, \tag{9.1.12}$$

then we easily know that $E(t)$ is equivalent to $\sum_{i=1}^n E_i(t)$, i.e.,

$$E(t) \sim \sum_{i=1}^n E_i(t). \tag{9.1.13}$$

By a straightforward calculation under assumptions $(A_1)-(A_2)$ (see below), we have

$$\frac{dE(t)}{dt} = \frac{1}{2}(g' \circ \nabla u) - \sum_{i=0}^n \left(\frac{1}{2} g^i(t) \int_\Omega | \nabla u^i |^2 dx \right)$$
$$- \int_\Omega | \nabla\theta_t |^2 dx - \int_{\Gamma_1} u_t \cdot H(u_t) d\Gamma \le 0, \tag{9.1.14}$$

where

$$u_t \cdot H(u_t) = \sum_{i=1}^n u_t^i h^i(u_t^i) \ge 0. \tag{9.1.15}$$

This indicates that the energy $E(u,\theta)(t)$ decreases on $(0, +\infty)$. However, the most interesting question is whether the energy decays exponentially or polynomially as $t \to +\infty$.

The aim of this chapter is to first extend the decay results in [173] for a viscoelastic system to those for the thermoviscoelastic system (9.1.8).

Throughout this chapter, we use "$\to 0$" to denote the strong convergence to zero in the corresponding space.We adopt the results in this chapter from [221].

9.2 Preliminaries

We shall consider the following hypotheses, let $\eta(t) = (\eta^1, \eta^2, \ldots, \eta^n)$. (A_1) $g^i{:}\mathbb{R}^+ \to \mathbb{R}^+$, $g^i(0) > 0$ $(i = 1, \ldots, n)$ is a differentiable function satisfying $(H_1)-(H_3)$ and that there exists a non-increasing differentiable function η^i satisfying

$$(g^i)'(t) \le -\eta^i(t)g^i(t),\ t \ge 0 \ \text{ and } \ \int_0^{+\infty} \eta^i(t)dt = +\infty, i = 1, \ldots, n.$$

(A_2) $H(u) = \left(h^1(u^1), \ldots, h^n(u^n)\right)$, let $H_0(|u|) = \left(h_0^1(|u^1|), \ldots, h_0^n(|u^n|)\right)$, and $h^i : \mathbb{R} \to \mathbb{R}$ $(i = 1, \ldots, n)$ are non-decreasing C^0 functions such that there exist strictly increasing functions $h_0^i \in C^1([0, +\infty))$ with $h_0^i(0) = 0$, and positive constants c_1^i, c_2^i, and ϵ^i such that

$$\begin{aligned} h_0^i(|s|) \le | h^i(s) | \le (h_0^i)^{-1}(|s|) \quad &\text{for all} \quad | s | \le \epsilon^i, \quad i = 1, \ldots, n, \\ c_1^i|s| \le | h^i(s) | \le c_2^i|s| \quad &\text{for all} \quad | s | \ge \epsilon^i, \quad i = 1, \ldots, n. \end{aligned}$$

Hypothesis (A_2) implies that $sh^i(s) > 0$, for all $s \ne 0$, $i = 1, 2, \ldots n$.

We set

$$V = \left\{ v \in H^1(\Omega) : v = 0 \text{ on } \Gamma_0 \right\}$$

and define the norm of V by

$$\| v \|_V = \left\{ \frac{1}{2} \| \nabla v \|^2_{L^2(0,l)} + \frac{1}{2} \int_{\Gamma_1} u \cdot H(u) d\Gamma \right\}^{\frac{1}{2}} \tag{9.2.1}$$

for any $H(u)$ satisfying (A_2). The treatment of well-posedness of system (9.1.8) is standard. For the sake of completeness, we give a brief discussion by using the theory of semigroups.

In order to use the semigroup approach, we introduce the new variable

$$w^t(x, s) = u(x, t) - u(x, t - s), \tag{9.2.2}$$

i.e.,

$$w^{t,i}(x, s) = u^i(x, t) - u^i(x, t - s),$$

consequently we obtain the following initial and boundary conditions

$$\begin{cases} w^t(x,0)=0, & for\ \ all\ t\geq 0,\\ w^t(x,s)=0, & (x,t)\in\Gamma_0\times(0,+\infty),\\ w^0(x,s)=w_0(x,s). \end{cases}$$

On the other hand, we may extend the solution u to negative times, setting $u(t)=0$ for $t<0$ and $v=u_t$. Then system (9.1.8) can be transformed into the system

$$\begin{cases} u_t-v=0, & (x,t)\in\Omega\times(0,+\infty),\\ v_t-l\Delta u-\int_0^{+\infty}g(s)\Delta w^t(t-s)ds+\nabla\theta_t=0, & (x,t)\in\Omega\times(0,+\infty),\\ \theta_{tt}-\Delta\theta_t-\Delta\theta+\operatorname{div}u_t=0, & (x,t)\in\Omega\times(0,+\infty),\\ w_t^t=-w_s^t+u_t, & (x,t)\in\Omega\times(0,+\infty),\\ \theta=0, & (x,t)\in\partial\Omega\times(0,+\infty),\\ u=0, & (x,t)\in\Gamma_0\times(0,+\infty),\\ l\frac{\partial u}{\partial\nu}+\int_0^t g(t-\tau)\frac{\partial w^t}{\partial\nu}d\tau+H(u_t)=0, & (x,t)\in\Gamma_1\times(0,+\infty), \end{cases} \tag{9.2.3}$$

where g satisfies (A_1),

$$\int_0^{+\infty}g(s)\Delta w^t(t-s)ds=\left(\int_0^{+\infty}g^1(s)\Delta w^{t,1}(t-s)ds,\ldots,\int_0^{+\infty}g^n(s)\Delta w^{t,n}(t-s)ds\right).$$

In view of (A_1), let $L_g^2(\mathbb{R}^+,V)$ be the Hilbert space of V-valued functions on $\mathbb{R}^+$, endowed with the norm

$$\| w^t \|^2_{(L_g^2(\mathbb{R}^+,V))^n}=\int_0^{+\infty}\sum_{i=1}^n g^i(s)\ \| w^{t,i}(s)\|_V^2\,ds.$$

We consider the problem in the following Hilbert space

$$\mathcal{H}=(V)^n\times(L^2)^n\times H_0^1\times L^2\times(L_g^2(\mathbb{R}^+,V))^n$$

with the energy norm

$$\|(u,v,\theta,\theta_t,w^t)\|_{\mathcal{H}}=\Big\{\sum_{i=1}^n l_i\|u^i\|_V^2+\frac{1}{2}\Big(\|v\|^2+\|\theta_t\|^2+\|\nabla\theta\|^2\Big)$$
$$+\int_0^{+\infty}\sum_{i=1}^n g^i(s)\ \| w^{t,i}(s)\|_V^2\,ds\Big\}^{\frac{1}{2}} \tag{9.2.4}$$

and define a linear unbounded operator A on $\mathcal{H}$ by

$$A(u,v,\theta,\theta_t,w^t)=(v,B(u,w^t)-\nabla\theta_t,\theta_t,\Delta\theta_t+\Delta\theta-\operatorname{div}v,v-w_s^t) \tag{9.2.5}$$

where $w_s^t = \frac{\partial w^t}{\partial s}$ and

$$B(u, w^t) = l \cdot \Delta u + \int_0^{+\infty} g(s) \cdot \Delta w^t(s)ds,$$

then problem (9.2.3) can be formulated as an abstract Cauchy problem

$$\Phi' = A\Phi, \quad \Phi = (u, v, \theta, \theta_t, w^t)$$

on the Hilbert space $\mathcal{H}$ for an initial data $\Phi(0) = (u_0, u_1, \theta_0, \theta_1, w_0^t)$. The domain of A is given by

$$\begin{aligned} D(A) = \Big\{ &(u, v, \theta, \theta_t, w) \in \mathcal{H} \ : \theta \in L^\infty\Big(\mathbb{R}^+, H_0^1(\Omega)\Big), \theta_t \in L^\infty\Big(\mathbb{R}^+, H_0^1(\Omega)\Big), \\ &\theta + \theta_t \in L^\infty\Big(\mathbb{R}^+, H^2(\Omega) \cap H_0^1(\Omega)\Big), v \in \Big(L^\infty(\mathbb{R}^+, V)\Big)^n, \\ &l_i u^i + \int_0^{+\infty} g^i(s) w^{t,i}(s)\, ds \in L^\infty\Big(\mathbb{R}^+, H^2(\Omega) \cap V\Big), \\ &w^t(s) \in \Big(L^\infty(\mathbb{R}^+, H_g^1(\mathbb{R}^+, V))\Big)^n, w^t(0) = 0, \\ &l_i \frac{\partial u^i}{\partial \nu} + \int_0^{+\infty} g^i(t-\tau) \frac{\partial w_t^{t,i}}{\partial \nu} d\tau + h^i(u_t^i) = 0 \quad \text{on } \Gamma_1, i = 1, \ldots, n \Big\}. \end{aligned} \tag{9.2.6}$$

It is clear that $D(A)$ is dense in $\mathcal{H}$.

We first prove that A is dissipative and closed. Indeed, by a straightforward calculation, it follows from Lemma 3.2 in [148] that for any $(u, v, \theta, \theta_t, w^t) \in D(A)$

$$\begin{aligned} &A(u, v, \theta, \theta_t, w^t), (u, v, \theta, \theta_t, w^t)\rangle_{\mathcal{H}} \\ &= (v, u)_V + \frac{1}{2}(B(u, w^t) - \nabla\theta_t, v) + \frac{1}{2}(\nabla\theta_t, \nabla\theta) \\ &\quad + \frac{1}{2}(\Delta\theta_t + \Delta\theta - (\mathrm{div} v, \theta) + (v - w_s^t, w^t)_{(L_g^2(\mathbb{R}^+, V))^n} \\ &= -\frac{1}{2}\int_{\Gamma_1} u \cdot H(u) d\Gamma - \frac{1}{2} \parallel \nabla\theta_t \parallel^2 - \sum_{i=1}^n g^i(s) \parallel w^{t,i}(s) \parallel_V^2 \Big|_0^{+\infty} \\ &\quad + \int_0^{+\infty} \sum_{i=1}^n (g^i)'(s) \parallel w^{t,i}(s) \parallel_{H_{\Gamma_1}^1(\Omega)}^2 ds \\ &\le 0, \end{aligned}$$

where $(\cdot, \cdot)$ stands for L^2-inner product on L^2. Thus, A is dissipative.

To prove that A is closed, let $(u_m, v_m, \theta_m, \theta_{mt}, w_m^t) \in D(A)$ be such that as $m \to +\infty$

$$(u_m, v_m, \theta_m, \theta_{mt}, w_m^t) \to (u, v, \theta, \theta_t, w^t) \quad \text{in } \mathcal{H}$$

and

$$A(u_m, v_m, \theta_m, \theta_{mt}, w_m^t) \to (a, b, c, d, e) \quad \text{in } \mathcal{H}.$$

Then we have

$$\begin{cases} u_m \to u \quad \text{strongly} \quad \text{in } \Big(L^\infty(\mathbb{R}^+, V\Big)^n, & (9.2.7) \\ v_m \to v \quad \text{strongly} \quad \text{in } \Big(L^\infty(\mathbb{R}^+, L^2(\Omega)\Big)^n, & (9.2.8) \\ \theta_m \to \theta \quad \text{strongly} \quad \text{in } L^\infty\Big(\mathbb{R}^+, H_0^1(\Omega)\Big), & (9.2.9) \\ \theta_{mt} \to \theta_t \quad \text{strongly} \quad \text{in } L^\infty\Big(\mathbb{R}^+, L^2(\Omega)\Big), & (9.2.10) \\ w_m^t \to w^t \quad \text{strongly} \quad \text{in } \Big(L^\infty(\mathbb{R}^+, L_g^2(\mathbb{R}^+, V)\Big)^n & (9.2.11) \end{cases}$$

and

$$\begin{cases} v_m \to a \quad \text{strongly} \quad \text{in } \Big(L^\infty(\mathbb{R}^+, V)\Big)^n, & (9.2.12) \\ B(u_m, w_m) - \nabla\theta_m \to b \quad \text{strongly} \quad \text{in } \Big(L^\infty(\mathbb{R}^+, L^2(\Omega))\Big)^n, & (9.2.13) \\ \theta_{mt} \to c \quad \text{strongly} \quad \text{in } L^\infty\Big(\mathbb{R}^+, H_0^1(\Omega)\Big), & (9.2.14) \\ \Delta\theta_{mt} + \Delta\theta_m - \operatorname{div} v_m \to d \quad \text{strongly} \quad \text{in } L^\infty\Big(\mathbb{R}^+, L^2(\Omega)\Big), & (9.2.15) \\ v_m - w_{ms}^t \to e \quad \text{strongly} \quad \text{in } \Big(L^\infty(\mathbb{R}^+, L_g^2(\mathbb{R}^+, V)\Big)^n. & (9.2.16) \end{cases}$$

From (9.2.8) and (9.2.12), we deduce

$$v_m \to v \quad \text{strongly} \quad \text{in } \Big(L^\infty(\mathbb{R}^+, V)\Big)^n \tag{9.2.17}$$

and

$$v = a \in \Big(L^\infty(\mathbb{R}^+, V)\Big)^n. \tag{9.2.18}$$

By (9.2.10) and (9.2.14), we get

$$\theta_{mt} \to \theta_t \quad \text{strongly} \quad \text{in } L^\infty\Big(\mathbb{R}^+, H_0^1(\Omega)\Big), \tag{9.2.19}$$

and

$$\theta_t = c \quad \text{in } L^\infty\Big(\mathbb{R}^+, H_0^1(\Omega)\Big). \tag{9.2.20}$$

Similarly, from (9.2.15) and (9.2.17), we derive

$$\Delta\theta_{mt} + \Delta\theta_m \to d + divv \quad \text{strongly} \quad \text{in } L^\infty\Big(\mathbb{R}^+, L^2(\Omega)\Big). \tag{9.2.21}$$

Thus, it follows from (9.2.9) and (9.2.19) that

$$\theta_{mt} + \theta_m \to \theta_t + \theta \quad \text{strongly} \quad \text{in } L^\infty\Big(\mathbb{R}^+, H^2(\Omega) \cap H_0^1(\Omega)\Big), \tag{9.2.22}$$

since Δ is an isomorphism from $H^2(\Omega) \cap H_0^1(\Omega)$ onto $L^2(\Omega)$. It therefore follows from (9.2.15) and (9.2.22) that

$$d = \Delta\theta_t + \Delta\theta - \text{div}v, \quad \theta_t + \theta \in L^\infty\Big(\mathbb{R}^+, H^2(\Omega) \cap H_0^1(\Omega)\Big). \tag{9.2.23}$$

By (9.2.11), (9.2.16) and (9.2.17), we conclude

$$w_m^t \to w^t \quad \text{strongly} \quad \text{in } \Big(L^\infty(\mathbb{R}^+, H_g^1(\mathbb{R}^+, V)\Big)^n \tag{9.2.24}$$

and

$$e = v - w_s^t, \quad w^t \in \Big(H_g^1(\mathbb{R}^+, V)\Big)^n, \; w^t(0) = 0. \tag{9.2.25}$$

In addition, from (9.2.7), (9.2.11) and (9.2.19), we have

$$B(u_m, w_m^t) - \nabla\theta_{mt} \to B(u, w^t) - \nabla\theta_t \quad \text{strongly} \quad \text{in } \Big(L^\infty(\mathbb{R}^+, L^2(\Omega))\Big)^n \tag{9.2.26}$$

in the sense of distribution. It therefore follows from (9.2.13) and (9.2.26) that

$$b = B(u, w^t) - \nabla\theta_t, \quad B(u, w^t) \in \Big(L^\infty(\mathbb{R}^+, L^2(\Omega))\Big)^n \tag{9.2.27}$$

and consequently,

$$l_i u^i + \int_0^{+\infty} g^i(s) w^{t,i}(s) ds \in L^\infty(\mathbb{R}^+, H^2(\Omega) \cap V). \tag{9.2.28}$$

Moreover, by (9.2.17), (9.2.28) and the trace theorem, we deduce

$$l_i \frac{\partial u^i}{\partial \nu} + \int_0^{+\infty} g^i(t-\tau) \frac{\partial w^{t,i}}{\partial \nu} d\tau + h^i(u_t^i) = 0 \quad \text{on } \Gamma_1, \quad i = 1, \ldots, n.$$

Thus, from (9.2.18), (9.2.20), (9.2.23), (9.2.25), (9.2.27) and (9.2.28), it follows

$$A(u, v, \theta, \theta_t, w^t) = (a, b, c, d, e), \quad (u, v, \theta, \theta_t, w^t) \in D(A).$$

Hence A is closed.

Next, we prove that $0 \in \rho(A)$, where $\rho(A)$ is the resolvent of the operator A. For any $G = (f_1, f_2, f_3, f_4, f_5) \in \mathcal{H}$, consider

$$A\Phi = G, \tag{9.2.29}$$

i.e.,

$$\begin{cases} v = f_1, & \text{in } \left(L^\infty(\mathbb{R}^+, V)\right)^n, & (9.2.30) \\ B(u, w^t) - \nabla\theta = f_2, & \text{in } \left(L^\infty(\mathbb{R}^+, L^2(\Omega))\right)^n, & (9.2.31) \\ \theta_t = f_3, & \text{in } L^\infty\left(\mathbb{R}^+, L^2(\Omega)\right), & (9.2.32) \\ \Delta\theta_t + \Delta\theta - \operatorname{div} v = f_4, & \text{in } L^\infty(\mathbb{R}^+, L^2(\Omega)), & (9.2.33) \\ v - w_s^t = f_5, & \text{in } \left(L^\infty(\mathbb{R}^+, L_g^2(\mathbb{R}^+, V))\right)^n. & (9.2.34) \end{cases}$$

Inserting $v = f_1$ in (9.2.30) and $\theta_t = f_3$ in (9.2.32) into (9.2.33), we obtain

$$\Delta\theta = f_4 + \operatorname{div} f_1 - \Delta f_3 \in L^\infty\left(\mathbb{R}^+, L^2(\Omega)\right). \tag{9.2.35}$$

By the standard theory for the linear elliptic equations, we conclude $\theta \in L^\infty\left(\mathbb{R}^+, H^2(\Omega) \cap H_0^1(\Omega)\right)$ which is unique and satisfies (9.2.35).

We plug $v = f_1$ obtained from (9.2.30) into (9.2.34) to get

$$w_s^t = f_1 - f_5 \in \left(L^\infty(\mathbb{R}^+, L_g^2(\mathbb{R}^+, V))\right)^n. \tag{9.2.36}$$

Applying the standard theory for the linear elliptic equations again, we have a unique $w^t \in \left(L^\infty(\mathbb{R}^+, H_g^1(\mathbb{R}^+, V))\right)^n$ satisfying (9.2.36). Then plugging θ and w^t just obtained from solving (9.2.35), (9.2.36) respectively into (9.2.31) and applying the standard theory for the linear elliptic equations again yields the unique solvability of $u \in D(A)$ for (9.2.31), and such that $l_i u^i + \int_0^{+\infty} g^i(s) w^{t,i}(s)\, ds \in L^\infty\left(\mathbb{R}^+, H^2(\Omega) \cap V\right)$. Thus the unique solvability of (9.2.29) follows. It is clear from the regularity theory for the linear elliptic equations that $\|\Phi\|_{\mathcal{H}} \le K\|G\|_{\mathcal{H}}$ with K being a positive constant independent of Φ.

By Lemmas 1.3.17–1.3.18 (see, e.g., [201, 287]), we can obtain the following existence results.

Lemma 9.2.1 *Assume that* $(A_1)-(A_2)$ *are satisfied. Then*

(i) For every initial data $(u_0, u_1, \theta_0, \theta_1) \in \mathcal{H}$, *problem (9.1.8) has a unique global mild solution satisfying*

$$(u(t), v(t), \theta(t), \theta_t(t)) \in C([0, +\infty), \mathcal{H}).$$

Moreover, we have for every $t \in [0, +\infty)$,

$$\| (u(t), v(t), \theta(t), \theta_t(t)) \|_{\mathcal{H}} \leq C \| (u_0, v_0, \theta_0, \theta_1) \|_{\mathcal{H}} .$$

(ii) For every initial data $(u_0, u_1, \theta_0, \theta_1) \in D(A)$, *satisfying the compatibility condition*

$$\frac{\partial u_0^i}{\partial \nu} + h^i(u_1^i) = 0 \quad \text{on} \quad \Gamma_1, \quad i = 1, 2, \ldots, n,$$

then problem (9.1.8) has a unique global classical solution satisfying

$$(u(t), v(t), \theta(t), \theta_t(t)) \in C([0, +\infty), D(A)).$$

9.3 Main Results

In this section, we shall state and prove our main result. Theorem 1.2.8 will be of essential use in establishing our main result. First we establish several lemmas.

Lemma 9.3.1 *There exists a constant* $c > 0$ *such that*

$$\int_\Omega \sum_{i=1}^n \Big(\int_0^t g^i(t-s)(u^i(t) - u^i(s))ds \Big)^2 dx \leq c(g \circ \nabla u)(t)$$

for all $u \in V$.

Proof Using the Cauchy-Schwartz and Poincaré's inequalities, we can obtain

$$\begin{aligned}
&\int_\Omega \sum_{i=1}^n \Big(\int_0^t g^i(t-s)(u^i(t) - u^i(s))ds \Big)^2 dx \\
&= \sum_{i=1}^n \int_\Omega \Big(\int_0^t g^i(t-s)(u^i(t) - u^i(s))ds \Big)^2 dx
\end{aligned}$$

$$\le \sum_{i=1}^{n}\Big(\int_\Omega \int_0^t g^i(t-s)|u^i(t)-u^i(s)|^2 ds\Big)dx$$
$$\le c\sum_{i=1}^{n}\Big(\int_\Omega \int_0^t g^i(t-s)|\nabla u^i(t)-\nabla u^i(s)|^2 ds\Big)dx = c(g\circ\nabla u)(t).$$

□

Now we are going to construct a Lyapunov functional $\mathcal{F}(t)$ equivalent to $E(t)$. To this end, we define several functionals which allow us to obtain the needed estimates.

Lemma 9.3.2 *Under the assumptions* $(A_1)-(A_2)$*, the function* F_1 *defined by*

$$F_1(t) = \int_\Omega u\cdot u_t dx$$

satisfies the estimate

$$\begin{aligned} F_1'(t) \le & -\frac{nl}{2}\int_\Omega |\nabla u|^2 dx + \int_\Omega |u_t|^2 dx + c(g\circ\nabla u)(t) \\ & + \frac{\lambda_0^2}{4\varepsilon}\int_\Omega |\nabla\theta_t|^2 dx + c\int_{\Gamma_1} |H(u_t)|^2 d\Gamma, \end{aligned} \tag{9.3.1}$$

where λ_0 *is the best constant in Poincaré's inequality, and* $l = \min(l_1,\ldots,l_n)$.

Proof Differentiating $F_1(t)$ with respect to t and using (9.1.8), we obtain

$$\begin{aligned} F_1'(t) = & \int_\Omega |u_t|^2 dx + \int_\Omega u\cdot\Delta u dx - \int_\Omega u\cdot\nabla\theta_t dx \\ & - \int_\Omega u\cdot\Big(\int_0^t g(t-\tau)\Delta u(\tau)d\tau\Big)dx \\ \le & \int_\Omega |u_t|^2 dx - \sum_{i=1}^{n} l_i\int_\Omega |\nabla u^i|^2 dx - \int_\Omega u\cdot\nabla\theta_t dx \\ & + \sum_{i=1}^{n}\Big(\int_\Omega \nabla u^i\Big(\int_0^t g^i(t-\tau)(\nabla u^i(\tau)-\nabla u^i(t))d\tau\Big)dx\Big) \\ & - \int_{\Gamma_1} u\cdot H(u_t)d\Gamma. \end{aligned} \tag{9.3.2}$$

Using Young's inequality and Lemma 9.2.1, we obtain for any $\varepsilon > 0$,

$$\sum_{i=1}^{n}\Big(\int_\Omega \nabla u^i\Big(\int_0^t g^i(t-\tau)(\nabla u^i(\tau)-\nabla u^i(t))d\tau\Big)dx\Big)$$

$$\leq \varepsilon \int_\Omega |\nabla u|^2 dx + \frac{1}{4\varepsilon} \sum_{i=1}^n \Big(\int_\Omega \Big(\int_0^t g^i(t-\tau) \mid \nabla u^i(\tau) - \nabla u^i(t) \mid d\tau \Big)^2 dx \Big)$$

$$\leq \varepsilon \int_\Omega |\nabla u|^2 dx + \frac{c}{\varepsilon}(g \circ \nabla u)(t). \tag{9.3.3}$$

Using Poincaré's and Young's inequalities, we find that for any $\varepsilon > 0$,

$$- \int_\Omega u \cdot \nabla \theta_t \leq \varepsilon \int_\Omega |\nabla u|^2 dx + \frac{\lambda_0^2}{4\varepsilon} \int_\Omega |\nabla \theta_t|^2 dx. \tag{9.3.4}$$

Similarly, using the Trace Theorem gives

$$- \int_{\Gamma_1} u \cdot H(u_t) d\Gamma \leq \varepsilon \int_{\Gamma_1} |\nabla u|^2 d\Gamma + \frac{\lambda_0^2}{4\varepsilon} \int_{\Gamma_1} |H(u_t)|^2 d\Gamma. \tag{9.3.5}$$

Combining (9.3.3)–(9.3.5) and choosing $\varepsilon > 0$ small enough, (9.3.1) follows. □

Lemma 9.3.3 *Under the assumptions* $(A_1)-(A_2)$, *the function* F_2 *defined by*

$$F_2(t) = - \sum_{i=1}^n \int_\Omega u_t^i \cdot \Big(\int_0^t g^i(t-\tau)(u^i(t) - u^i(\tau)) d\tau \Big) dx$$

satisfies the estimate

$$\begin{aligned} F_2'(t) \leq & - \sum_{i=1}^n \Big(\int_0^t g^i(s) ds - \varepsilon \Big) \int_\Omega |u_t^i|^2 dx + \varepsilon \int_\Omega (|\nabla u|^2 + |\nabla \theta_t|^2) dx \\ & + \frac{c}{\varepsilon}(g \circ \nabla u)(t) - \frac{c}{\varepsilon}(g' \circ \nabla u)(t) + c \int_{\Gamma_1} |H(u_t)|^2 d\Gamma \end{aligned} \tag{9.3.6}$$

for any $0 < \varepsilon < 1$.

Proof By exploiting (9.1.8) and integrating by parts, we have

$$\begin{aligned} F_2'(t) = & \sum_{i=1}^n \Big(\Big(1 - \int_0^t g^i(s) ds \Big) \int_\Omega \nabla u^i \cdot \Big(\int_0^t g^i(t-\tau)(\nabla u^i(t) - \nabla u^i(\tau)) d\tau \Big) dx \Big) \\ & + \sum_{i=1}^n \int_{\Gamma_1} \Big(\int_0^t g^i(t-\tau)(u^i(t) - u^i(\tau)) d\tau \Big) \cdot h^i(u_t^i) d\Gamma \\ & - \sum_{i=1}^n \int_\Omega u_t^i \cdot \Big(\int_0^t (g^i)'(t-\tau)(u^i(t) - u^i(\tau)) d\tau \Big) dx \\ & + \int_\Omega \nabla \theta_t \Big(\int_0^t g(t-\tau)(u(t) - u(\tau)) d\tau \Big) dx \end{aligned}$$

$$-\sum_{i=1}^{n}\Big(\int_0^t g^i(s)ds\Big)\int_\Omega |\, u_t^i \,|^2\, dx$$
$$+\sum_{i=1}^{n}\int_\Omega \Big(\int_0^t g^i(t-\tau)(\nabla u^i(t)-\nabla u^i(\tau))d\tau\Big)^2 dx.$$

Using Young's and Poincaré's inequalities and Lemma 9.3.1, we obtain for any $\varepsilon > 0$,

$$\sum_{i=1}^{n}\Big(\Big(1-\int_0^t g^i(s)ds\Big)\int_\Omega \nabla u^i\cdot\Big(\int_0^t g^i(t-\tau)(\nabla u^i(t)-\nabla u^i(\tau))d\tau\Big)dx\Big)$$
$$\leq \varepsilon\int_\Omega |\, \nabla u \,|^2\, dx + \frac{c}{\varepsilon}(g\circ\nabla u)(t),$$

$$\sum_{i=1}^{n}\int_{\Gamma_1}\Big(\int_0^t g^i(t-\tau)(u^i(t)-u^i(\tau))d\tau\Big)\cdot h(u_t^i)d\Gamma$$
$$\leq c(g\circ\nabla u)(t) + c\int_{\Gamma_1} |\, H(u_t)\,|^2\, d\Gamma,$$
$$-\sum_{i=1}^{n}\int_\Omega u_t^i\cdot\Big(\int_0^t (g^i)'(t-\tau)(u^i(t)-u^i(\tau))d\tau\Big)dx$$
$$\leq \varepsilon\int_\Omega |\, u_t\,|^2\, dx - \frac{\lambda_0^2}{4\varepsilon}(g'\circ\nabla u)(t), \int_\Omega \nabla\theta_t\Big(\int_0^t g(t-\tau)(u(t)-u(\tau))d\tau\Big)dx$$
$$\leq \varepsilon\int_\Omega |\, \nabla\theta_t\,|^2\, dx + \frac{\lambda_0^2}{4\varepsilon}(g\circ\nabla u)(t).$$

Combining all above estimates, we can prove (9.3.6). □

Now, for $N_1, N_2 > 1$, let

$$\mathcal{F}(t) = N_1E(t) + N_2F_2(t) + F_1(t)$$

and set $g_0^i = \int_0^{t_0} g^i(s)ds > 0,\ g_0 = \min(g^1, \ldots, g^n)$ for some fixed $t_0 > 0$. Combining (9.1.12), (9.3.1) and (9.3.6), taking $\varepsilon > 0$ small enough, and N_1, N_2 large enough, we arrive at

$$\mathcal{F}'(t) \leq -\frac{nl}{4}\int_\Omega |\, \nabla u\,|^2\, dx - k\int_\Omega |\, u_t\,|^2\, dx - k_1\int_\Omega |\, \nabla\theta_t\,|^2\, dx + c(g\circ\nabla u)(t)$$
$$+c\int_{\Gamma_1} |\, H(u_t)\,|^2\, d\Gamma,$$

where

$$k = \Big(N_2 n g_0 - \frac{nl}{4} - n\Big) > 0, \quad k_1 = \Big(N_1 - \varepsilon N_2\Big) > 0,$$

which yields

$$\mathcal{F}'(t) \le -\beta_1 E(t) + c(g \circ \nabla u)(t) + c \int_{\Gamma_1} | H(u_t) |^2 \, d\Gamma, \quad \text{for} \quad t \ge t_0 \tag{9.3.7}$$

or

$$E(t) \le -\beta_2 \mathcal{F}'(t) + c(g \circ \nabla u)(t) + c \int_{\Gamma_1} | H(u_t) |^2 \, d\Gamma, \quad \text{for} \quad t \ge t_0. \tag{9.3.8}$$

On the other hand, we can choose N_1 so large that $\mathcal{F}(t)$ is equivalent to $E(t)$, i.e.,

$$\mathcal{F}(t) \sim E(t). \tag{9.3.9}$$

We now prove our main result.

Theorem 9.3.1 *Assume that* $(A_1)-(A_2)$ *hold. Then there exists a constant* $C > 0$ *such that, for large t, the solution of problem (9.1.8) satisfies*

$$E(t) \le C \sum_{i=1}^{n} \left((I_0^i)^{-1} \left(\frac{1}{\int_0^t \eta^i(s) ds} \right) \right)^2, \quad \text{where} \quad I_0(s) = sH_0(s). \tag{9.3.10}$$

Moreover, if K defined by $K(s) = \frac{H_0(s)}{s}$ *is strictly increasing with* $K(0) = 0$*, then we have the following improved estimate*

$$E(t) \le C \sum_{i=1}^{n} \left(h_0^{-1} \left(\frac{1}{\int_0^t \eta^i(s) ds} \right) \right)^2. \tag{9.3.11}$$

Proof Since the proof is similar to that of [173] for a viscoelastic system, we sketch the proof here. Define $\phi(t) = 1 + \int_1^t \frac{1}{H_0(1/s)} ds$, i.e., $\phi^i(t) = 1 + \int_1^t \frac{1}{h_0^i(1/s)} ds$. Then

$$(\phi^i)'(t) = \frac{1}{h_0^i(1/t)} > 0, \quad \textit{for all } t \ge 1, \quad (\phi^i)'(t) \to +\infty \quad \text{as} \quad t \to +\infty,$$

and $(\phi^i)'(t)$ is strictly increasing. Thus ϕ^i is convex and strictly increasing C^2 function, with $\phi^i(t) \to +\infty$ as $t \to +\infty$. If we set $\sigma_0^i = (\phi^i)^{-1}$, then σ_0^i is strictly increasing, $(\sigma_0^i)'(t) = h_0^i(1/t)$ is decreasing, and $(\sigma_0)^i(t) \to +\infty$ as $t \to +\infty$.

Now we define for all $t \ge t_1$,

$$\sigma(t) := \sigma_0\Big(\int_0^t \eta(s)ds\Big),$$

for some $t_1 \geq t_0$ with $\int_0^{t_1} \eta^i(s)ds \geq 1, \quad i = 1, 2, \ldots, n$. Using the properties of $(\sigma_0)^i$ and η^i, it is easy to check that σ^i is a strictly increasing and concave twice differentiable function, with $\sigma^i(t) \to +\infty$ as $t \to +\infty$. From (9.3.8)–(9.3.9), we have for $T \geq s \geq t_1, \quad i = 1, 2, \ldots, n$,

$$\begin{aligned}
\int_s^T (\sigma^i)'(t)E_i^2(t)dt \ &\leq c\int_s^T (\sigma^i)'E_i\mathcal{F}_i dt \leq -c_1\int_s^T (\sigma^i)'\mathcal{F}_i(\mathcal{F}_i)'dt \\
&\quad + c_2\int_s^T (\sigma^i)'\mathcal{F}_i(g^i \circ \nabla u^i)dt \\
&\quad + c_2\int_s^T (\sigma^i)'\mathcal{F}_i\Big(\int_{\Gamma_1} (h^i)^2(u_t^i)d\Gamma\Big)dt \\
&\leq cE_i^2(s) + c\int_s^T (\sigma^i)'E_i(g^i \circ \nabla u^i)dt \\
&\quad + c\int_s^T (\sigma^i)'E_i\Big(\int_{\Gamma_1} (h^i)^2(u_t^i)d\Gamma\Big)dt.
\end{aligned} \tag{9.3.12}$$

Now we estimate the last two terms of (9.3.12) for any fixed $i = 1, 2, \ldots, n$ as follows

$$I_1^i = \int_s^T (\sigma^i)'E_i(g^i \circ \nabla u^i)dt, \quad I_2^i = \int_s^T (\sigma^i)'E_i\Big(\int_{\Gamma_1} (h^i)^2(u_t^i)d\Gamma\Big)dt.$$

Using (A_1), and the definition of σ^i, and noting the fact that $(\sigma_0^i)'$ and η^i are non-increasing, we get

$$\begin{aligned}
(\sigma^i)'(t)g^i(t-\tau) = (\sigma_0^i)'\Big(\int_0^t \eta^i(s)ds\Big)\eta^i(t)g^i(t-\tau) &\leq c\eta^i(t)g^i(t-\tau) \\
&\leq c\eta^i(t-\tau)g^i(t-\tau) \leq -c(g^i)'(t-\tau).
\end{aligned}$$

Hence

$$I_1^i \leq -c\int_s^T E_i((g^i)' \circ \nabla u^i)dt \leq -c\int_s^T E_iE_i'dt \leq cE_i^2(s). \tag{9.3.13}$$

For I_2^i, we consider the following partition of Γ_1,

$$\begin{cases} \Gamma_{11}^i = \{x \in \Gamma_1 :\mid u_t^i \mid > \epsilon\}, \\ \Gamma_{12}^i = \Big\{x \in \Gamma_1 :\mid u_t^i \mid \le \epsilon \text{ and } \mid u_t^i \mid \le (\sigma_0^i)'\Big(\int_0^t \eta^i(s)ds\Big)\Big\}, \\ \Gamma_{13}^i = \Big\{x \in \Gamma_1 :\mid u_t^i \mid \le \epsilon \text{ and } \mid u_t^i \mid > (\sigma_0^i)'\Big(\int_0^t \eta^i(s)ds\Big)\Big\}. \end{cases} \tag{9.3.14}$$

Using (A_2), (9.1.12), and the properties of σ^i, σ_0^i and η^i, we arrive at

$$(\sigma^i)'(t)\int_{\Gamma_{11}^i}(h^i)^2(u_t^i)d\Gamma \le c(\sigma^i)'(t)\int_{\Gamma_{11}^i}u_t^i h^i(u_t^i)d\Gamma \le -cE_i'(t), \tag{9.3.15}$$

$$\begin{aligned}(\sigma^i)'(t)\int_{\Gamma_{12}^i}(h^i)^2(u_t^i)d\Gamma &\le \eta^i(t)(\sigma_0^i)'\Big(\int_0^t \eta^i(s)ds\Big)\int_{\Gamma_{12}^i}\big((h_0^i)^{-1}(|u_t^i|)\big)^2 d\Gamma \\ &\le \eta^i(t)(\sigma_0^i)'\Big(\int_0^t \eta^i(s)ds\Big)\Big((h_0^i)^{-1}\Big((\sigma_0^i)'\Big(\int_0^t \eta^i(s)ds\Big)\Big)\Big)^2,\end{aligned} \tag{9.3.16}$$

$$\begin{aligned}(\sigma^i)'(t)\int_{\Gamma_{13}^i}(h^i)^2(u_t^i)d\Gamma &= \eta^i(t)(\sigma_0^i)'\Big(\int_0^t \eta^i(s)ds\Big)\int_{\Gamma_{13}^i}(h^i)^2(u_t^i)d\Gamma \\ &\le \eta^i(t)(h_0^i)^{-1}(\epsilon)\int_{\Gamma_{13}^i}u_t^i h^i(u_t^i)d\Gamma \le -cE_i'(t).\end{aligned} \tag{9.3.17}$$

Thus it follows from (9.3.15)–(9.3.17),

$$\begin{aligned}&(\sigma^i)'(t)\int_{\Gamma_1^i}(h^i)^2(u_t^i)d\Gamma \\ &\quad\le -cE_i'(t) + c\eta^i(t)(\sigma_0^i)'\Big(\int_0^t \eta^i(s)ds\Big)\Big((h_0^i)^{-1}\Big((\sigma_0^i)'\Big(\int_0^t \eta^i(s)ds\Big)\Big)\Big)^2\end{aligned}$$

which implies

$$I_2^i \le cE_i^2(s) + cE_i(s)\int_s^T \eta^i(t)(\sigma_0^i)'\Big(\int_0^t \eta^i(s)ds\Big)\Big((h_0^i)^{-1}\Big((\sigma_0^i)'\Big(\int_0^t \eta^i(s)ds\Big)\Big)\Big)^2 dt. \tag{9.3.18}$$

Therefore, a combination of (9.3.12)–(9.3.13) with (9.3.18) yields

$$\begin{aligned}&\int_s^T (\sigma^i)'(t)E_i^2(t)dt \\ &\quad\le cE_i^2(s) + cE_i(s)\int_s^{+\infty}\eta^i(t)(\sigma_0^i)'\Big(\int_0^t \eta^i(s)ds\Big)\Big((h_0^i)^{-1}\Big((\sigma_0^i)'\Big(\int_0^t \eta^i(s)ds\Big)\Big)\Big)^2 dt \\ &\quad= cE_i^2(s) + cE_i(s)\int_{\int_0^s \eta^i(s)ds}^{+\infty}(\sigma_0^i)'(\tau)((h_0^i)^{-1}((\sigma_0^i)'(\tau)))^2 d\tau \\ &\quad= cE_i^2(s) + cE_i(s)\int_{(\sigma_0^i)(\int_0^s \eta^i(s)ds)}^{+\infty}\Big((h_0^i)^{-1}\Big((h_0^i)(1/s)\Big)\Big)^2 ds \\ &\quad= cE_i^2(s) + \frac{cE_i(s)}{(\sigma_0^i)(\int_0^s \eta^i(s)ds)} = cE_i^2(s) + \frac{cE_i(s)}{\sigma^i(s)}.\end{aligned}$$

Thus using Theorem 1.2.8 leads to

$$E_i(t) \le \frac{c}{(\sigma^i(t))^2} = \frac{c}{\Big((\sigma_0^i)(\int_0^s \eta^i(s)ds)\Big)^2}, \quad for \;\; all \; t \ge t_1. \tag{9.3.19}$$

To obtain (9.3.10), we take s_0 such that $h_0^i(1/s_0) \le 1$. Since h_0 is increasing and $I_0(s) = sH_0(s)$, we get

$$(\sigma_0^i)^{-1}(s) \le 1 + (s-1)\frac{1}{h_0^i(1/s)} \le \frac{s}{h_0^i(1/s)} = \frac{1}{I_0^i(1/s)}, \forall s \ge s_0,$$

which, together with $t = \frac{1}{I_0^i(1/s)}$, yields

$$\frac{1}{\sigma_0(t)} \le (I_0^i)^{-1}(1/t), \quad for \;\; all \;\; t \ge 1.$$

Noting (9.1.12), this yields (9.3.10) by virtue of (9.3.19).

To prove (9.3.11), we take

$$\sigma_0^i = (\phi^i)^{-1}, \quad \text{where} \quad \phi^i(t) = 1 + \int_1^t \frac{1}{K^i(1/s)} ds, \forall t \ge 1,$$

and replace (9.3.14) by

$$\begin{cases} \Gamma_{11}^i = \{x \in \Gamma_1^i : \mid u_t^i \mid > \epsilon\}, \\ \Gamma_{12}^i = \Big\{x \in \Gamma_1^i : \mid u_t^i \mid \le \epsilon \text{ and } (h_0^i)^{-1}(\mid u_t^i \mid) \le (K^i)^{-1}\Big((\sigma_0^i)'\Big(\int_0^t \eta^i(s)ds\Big)\Big)\Big\}, \\ \Gamma_{13}^i = \Big\{x \in \Gamma_1^i : \mid u_t^i \mid \le \epsilon \text{ and } (h_0^i)^{-1}(\mid u_t^i \mid) > (K^i)^{-1}\Big((\sigma_0^i)'\Big(\int_0^t \eta^i(s)ds\Big)\Big)\Big\}. \end{cases}$$

Then repeating the same procedure of the proof of (9.3.10), we can establish (9.3.11). □

Similarly to [173], we shall give some examples to illustrate the energy decay rates obtained from Theorem 9.3.1.

(1) Between polynomial and exponential growth ($n = 1$).

If $h_0(s) = e^{-(\ln s)^2}$ near zero, then we have the energy decay rate (in this case $\eta \ne \eta_1$)

$$E(t) \le Ce^{-2(\ln(\int_0^t \eta(s)ds))^{1/2}}.$$

(2) Exponential growth ($n = 1$).

If $h_0(s) = e^{-1/s}$ near zero, then we have the energy decay rate (in this case $\eta \ne \eta_1$)

$$E(t) \leq \frac{C}{\left(\ln(\int_0^t \eta(s)ds)\right)^2}.$$

(3) Faster than exponential growth ($n = 1$).

If $h_0(s) = e^{-e^{1/s}}$ near zero, then we have the energy decay rate (in this case $\eta = \eta_1$)

$$E(t) \leq \frac{C}{\left(\ln\left(\mid \ln(\int_0^t \eta(s)ds) \mid\right)\right)^2}.$$

9.4 Bibliographic Comments

If the relaxation function $g(t)$ satisfies conditions $(H_1) - (H_3)$, Navarro [196] considered the following model

$$\begin{cases} u_{tt} - \mu\Delta u - (\lambda + \mu)\nabla \text{div} u \\ \quad +\mu g * \Delta u + (\lambda + \mu) g * \nabla \text{div} u + \alpha \nabla\theta = 0, \ (x, t) \in \Omega \times (0, +\infty), \\ \theta_t - \Delta\theta + \beta \text{div} u_t = 0, \qquad (x, t) \in \Omega \times (0, +\infty), \end{cases} \tag{9.4.1}$$

and proved the asymptotic stability for system (9.4.1), that is,

$$\lim_{t\to+\infty} \| u(t) \|_{(H_0^1(\Omega))^n} = \lim_{t\to+\infty} \| u'(t) \|_{(L^2(\Omega))^n} = \lim_{t\to+\infty} \| \theta(t) \|_{L^1(\Omega)} = 0.$$

Later on, Liu and Zheng [154] proved that the energy decays to zero exponentially; Liu [148] proved the exponential stability with boundary feedback, and obtained the explicit decay rate. For the thermoviscoelastic plates, Muñoz Rivera and Barreto [188] concerned with decay rates of solutions to the thermoviscoelastic plate equations of integral type. It is already known that solutions of the thermoelastic plate equations decay to zero exponentially as time goes to infinity, adding the viscoelastic damping of integral type to the thermoelastic plate equations, one should expect that solutions of the thermoviscoelastic plate equations continue to converge to zero exponentially even if the relaxation kernel decays polynomially, but this is not true, as an example in this chapter shows, since the memory effect prevails over thermal diffusion in the equations. More precisely, the authors proved that if the relaxation decays to zero exponentially or polynomially, the first order energy of the solutions converges to zero exponentially or polynomially too. The proof was based on the energy estimates and the construction of a new Lyapunov functional. Potomkin [203] considered a model of thermoviscoelastic plates, where thermal effects are given by the Gurtin and Pipkin law and the viscoelastic effects by the past history of the material. He also considered a nonlinearity given by the Berger model for plates. Using new history variables, the author transformed the original system into an evolutionary system. Thus, using the semigroup theory, he proved the well-posedness of the problem and also established explicit representation formulas for the new variables. Recently, El-Karamany and Ezzat [63] considered two-temperature Green-Naghdi theory of type III in linear thermoviscoelastic anisotropic solid, and in this paper, the reciprocal theorem is proved and the variational principle is established for the linear two-temperature Green-Naghdi theory of type III in an anisotropic and inhomogeneous thermoviscoelastic solid. A proof of a uniqueness theorem for thermoviscoelasticity, without restrictions imposed on the relaxation or thermal conductivity tensors, except symmetry conditions, is given. For the second sound, Berti et al. [16] dealt with a contact problem describing the mechanical and thermal evolution of a damped extensible thermoviscoelastic beam under the Cattaneo's law, relating the heat flux to the gradient of the temperature. The beam is rigidly clamped at its left end whereas the right end of the beam moves vertically between reactive stops like a nonlinear spring. In this chapter, existence and uniqueness of the solution is proved, as well as the exponential decay of the related energy. Then, fully discrete approximations are introduced by using the classical finite element method and the implicit Euler scheme to approximate the spatial variable and to discretize the time derivatives, respectively.

In the case of higher space dimensions, the problem is very complicated. In order to see such complexity, we look at some special cases: thermoelastic and viscoelastic system.

When $g = 0$, system (9.1.1) reduces to the thermoelastic system of type I without memory effects

$$\begin{cases} u_{tt} - \Delta u + \nabla\theta = 0, \\ \theta_t - \Delta\theta + \nabla u_t = 0. \end{cases}$$

The corresponding energy $E(u, \theta, t)$ is defined by

$$E(u, \theta, t) = \frac{1}{2}\int_{\Omega}\left(u_t^2 + \theta^2 + |\nabla u|^2\right)dx.$$

For the thermoelasticity of type I, there are many works (see, e.g., [47, 90, 110, 120, 155, 183, 184,185, 257, 274]) on the existence, uniqueness and asymptotic behavior of solutions of the linear system, for example, Slemrod [257] proved the global existence, uniqueness and asymptotic stability of classical smooth solutions; Shibata [255] considered the initial boundary value problem with the boundary conditions $u_x - \gamma\theta = 0, \theta_x = 0 \ (x = 0, l)$; Racke, Shibata and Zheng [240] obtained the global existence and uniqueness of solutions for the nonlinear thermoelastic system of type I with small initial data; Muñoz Rivera and Qin [189] proved the global existence, uniqueness, and asymptotic behavior of solutions for the one-dimensional nonlinear thermoelasticity with thermal memory subject to Dirichlet-Dirichlet boundary conditions.

When $g = 0$ and $h = 0$, system (9.1.1) reduces to the thermoelastic system of type III without memory effects and boundary feedback

$$\begin{cases} u_{tt} - \Delta u + \nabla\vartheta = 0, \\ \vartheta_{tt} - \Delta\vartheta_t - \Delta\vartheta + \nabla\cdot u_{tt} = 0. \end{cases}$$

This model comes from the theory of Green and Naghdi [89, 90, 91] who reexamined the classical Fourier's law in thermoelasticity and introduced the so-called models of thermoelasticity of types II and III, whose constitutive assumptions on the heat flux vector are different from the Fourier's law, based on which the thermoelastic model of type I was proposed.

There are some interesting results for the thermoelastic system of type III, Zhang and Zuazua [285], Quintanilla and Racke [233] independently studied the decay of energy for the problem of the linear thermoelastic system of type III by using the classical energy method and the spectral method, and obtained the exponential stability in one space dimension, and in two or three space dimensions for radially symmetric situations while the energy decays polynomially for most domains in two or three space dimensions. Recently, Messaoudi and Said-Houari [175] considered a Timoshenko-type system of thermoelasticity of type III, and proved an exponential decay result. The Cauchy problem of the linear thermoelastic system of type III was also studied by Reissig and Wang [243], Zhang and Zuazua [285], the asymptotic behavior of the solution was studied by Lazzari and Nibbi [140] for the three-dimensional thermoelastic systems of types II and III.

However, in the higher dimensional space case, it is well known that the energy, in general, does not tend to zero as $t \to +\infty$. Indeed, Lebeau and Zuazua [143] gave a sufficient and necessary condition ensuring that the energy tends to zero exponentially as $t \to +\infty$ in a bounded multi-dimensional smooth domain Ω. This condition is written in terms of the dynamics of the rays of geometric optics. As a consequence of the result of [143], it follows that when Ω is a bounded smooth convex open set, the energy does not decay exponentially to zero. This is because the total energy is not dissipated completely in the form of thermal energy. Therefore, in such case, in order to ensure the exponential stabilization, a boundary velocity feedback was introduced in Liu [147, 148], to increase the loss of energy for the model of thermoviscoelasticity.

When the system (9.1.1) reduces to the following viscoelastic system

$$u_{tt} - \Delta u + \int_0^t g(t-\tau)\Delta u(\tau)d\tau = 0,$$

the energy $E(t)$ is defined by

$$E(u,t) = \frac{1}{2}\int_\Omega \Big(u_t^2 + \Big(1 - \int_0^t g(s)ds\Big)|\nabla u|^2\Big)dx + \frac{1}{2}(g\circ\nabla u)(t).$$

There are many results (see, e.g., [15, 30, 57, 67, 68, 146, 154, 163, 179, 263]) on exponential stability of energy and asymptotic stability of solutions to viscoelastic systems under different assumptions. Giorgi and Vegni [86] investigated the longtime behavior of the mathematical model of a homogeneous viscoelastic plate based on Reissner-Mindlin deformation shear assumptions, they showed the behavior of the energy when time goes on. In particular, assuming that the kernels λ and μ decay exponentially, and not too weakly with respect to the physical properties considered in the model, then the energy decays uniformly with respect to the initial conditions. Messaoudi and Mustafa [173] considered the control of solutions for viscoelastic equations with boundary feedback. Recently, Araújo et al. [199] concerned a class of quasilinear wave equations with memory

$$| u_t |^\rho u_{tt} - \alpha \Delta u - \Delta u_{tt} + \int_\tau^t \mu(t-s)\Delta u(s)ds - \gamma \Delta u_t + f(u) = h, \ \rho > 0,$$

which was considered by several authors with $\tau = 0$, the authors provided some results on the well-posedness and long-time behavior to this equation in a more general setting which includes past history. Later on, the result was improved in [213], where the authors investigated the existence of uniform attractors for a non-autonomous viscoelastic equation with a past history by establishing the uniformly asymptotic compactness of the semi-process generated by the global solutions.

It can be easily seen from (9.1.12) that the asymptotic behavior of the energy $E(u, \theta, t)$ for problem (9.1.8) depends significantly on the relaxation function $g(t)$ and the history as well. Subsequently, extensive attention was paid to the problem of obtaining an explicit decay rate. In this aspect, Day [53] first obtained a decay rate of t^{-1} in the case of one space dimension by introducing a feedback at one end of an interval. Later on, in the case of two space dimension, Leugering [141] established an exponential decay rate by introducing a velocity feedback on a part of the boundary of domain. On the other hand, if $g(t)$ decays exponentially and the initial history is taken to be zero, Desch and Miller [56] proved that, in the case of one space dimension, the energy also decays to zero exponentially at a rate no better than $g(t)$ decays. Recently, Mustafa [193] considered an n-dimensional thermoelastic system of type III with viscoelastic damping. They established an explicit and general decay rate result without imposing restrictive assumptions on the behavior of the relaxation function at infinity. Their result allows a larger class of relaxation functions and generalizes previous results existing in the literature. In view of the above statement, in order to obtain an explicit decay rate of energy of higher dimensional thermoelastic system (9.1.1), it may be indispensable to introduce a velocity feedback on a part of the boundary of a thermoelastic body. For the global attractors about thermoviscoelastic systems, we refer to Shang [250], and the authors of the present book studied the global existence, uniqueness and asymptotic behavior as time tends to infinity of the solution to a nonlinear one-dimensional thermoviscoelastic system. Furthermore, they investigated dynamics of the system and obtained the existence of global attractors.

We would like to mention other works in [45, 66, 101, 112, 261, 262] for related models.

References

1. Adams, R.A.: Sobolev Spaces. Academic Press, New York (1975)
2. Ait Ben Hassi, E.M., Ammari, K., Boulite, S., Maniar, L.: Stability of abstract thermo-elastic semigroups. J. Math. Anal. Appl. **435**, 1021–1035 (2016)
3. Alabau-Boussouira, F.: Asymptotic behavior for Timoshenko beams subject to a single nonlinear feedback control. Nonlinear Differ. Equ. Appl. **14**(4–5), 643–669 (2007)
4. Ammar-Khodja, F., Benabdallah, A.: Muñoz Rivera, J.E., Racke, R.: Energy decay for Timoshenko system of memory type. J. Differ. Equ. **194**(1), 82–115 (2003)
5. Aouadi, M., Lazzari, B., Nibbi, R.: A theory of thermoelasticity with diffusion under Green-Naghdi models. Zeitschrift for Angewandte and Mechanic **94**(10), 837–852 (2014)
6. Apalara, T.A., Messaoudi, S.A.: An exponential stability result of a Timoshenko system with thermoelasticity with second sound and in the presence of delay. Appl. Math. Optim. **71**, 449–472 (2015)
7. Araruna, F.D.: Braze Silva, P., Zuazua, E.: Asymptotic limits and stabilization for 1D nonlinear Mindlin-Timoshenko system. J. Syst. Sci. Complex **23**, 1–17 (2010)
8. Araújo, R.O., Ma, T.F., Qin, Y.: Long-time behavior of a quasilinear viscoelastic equation with past history. J. Differ. Equ. **254**, 4066–4087 (2013)
9. Avalos, G., Lasiecka, I.: Exponential stability of thermoelastic system without mechanical dissipation. Rend. Istit. Mat. Univ. Trieste Suppl. **XXVIII**, 1–28 (1997)
10. Avalos, G., Lasiecka, I.: Exponential stability of a thermoelastic system with free boundary conditions without mechanical dissipation. SIAM J. Math. Anal. **29**(1), 155–182 (1998)
11. Bahouri, H., Chemin, J.Y., Danchin, R.: Fourier Analysis and Nonlinear Partial Differential Equations. Springer, Berlin (2011)
12. Bassam, M., Mercier, D., Nicaisej, S., Wehbe, A.: Polynomial stability of the Timoshenko system by one boundary damping. J. Math. Anal. Appl. **425**, 1177–1203 (2015)
13. Batty, C.J.K.: Tauberian theorems for the Laplace-Stieltjes transform. Trans. Am. Math. Soc. **322**, 783–804 (1990)
14. Batty, C.J.K., Duyckaerts, T.: Non-uniform stability for bounded semi-groups on Banach spaces. J. Evol. Equ. **8**, 765–780 (2008)
15. Belleni-Morante, A., McBride, A.C.: Applied Nonlinear Semigroups. Wiley Series in Mathematical Mathods in Practice. Wiley, Chichester (1998)
16. Berrimi, S., Messaoudi, S.A.: Existence and decay of solutions of a viscoelastic equation with a nonlinear source. Nonlinear Anal. TMA **64**, 2314–2331 (2006)
17. Berti, A., Copetti, M.I.M., Fernandez, J.R., Naso, M.G.: A dynamic thermoviscoelastic contact problem with the second sound effect. J. Math. Anal. Appl. **421**, 1163–1195 (2015)

Y. Qin and Z. Ma, *Global Well-posedness and Asymptotic Behavior of the Solutions to Non-classical Thermo(visco)elastic Models*,
DOI 10.1007/978-981-10-1714-8

18. Bielak, J., MacCamy, R.C.: Dissipative boundary conditions for one-dimensional wave propagation. J. Integral Equ. Appl. **2**, 307–331 (1990)
19. Bonche, P., Koonin, S., Negele, J.W.: One-dimensional nuclear dynamics in the TDHF approximation. Phys. Rev. C **13**(3), 1226–1258 (1976)
20. Borichev, A., Tomilov, Y.: Optimal polynomial decay of functions and operator semigroups. Math. Ann. **347**, 455–478 (2010)
21. Bosello, C.A., Lazzari, B., Nibbi, R.: A viscous boundary condition with memory in linear elasticity. Int. J. Eng. Sci. **45**, 94–110 (2007)
22. Bresse, J.A.C.: Cours de Mécanique Appliquée, Mallet Bachelier (1859)
23. Brezis, H.: Opérateurs Maximaux Monotones et Semigroupes de Contractions dans le Espaces de Hilbert, Mathematics Studies, vol. 5. North-Holland, Amsterdam (1973)
24. Brokate, M., Sprekels, J.: Hysteresis and Phase Transitions. Springer, Heidelberg (1996)
25. Bucci, F., Chueshov, I.: Long-time dynamics of a coupled system of nonlinear wave and thermoelastic plate equations. Discrete Contin. Dyn. Syst. **22**, 557–586 (2008)
26. Burns, J.A., Liu, Z., Zheng, S.: On the energy decay of a linear thermoelastic bar. J. Math. Anal. Appl. **179**, 574–591 (1993)
27. Caraballo, T., Chueshov, I., Langa, J.: Existence of invariant manifolds for coupled parabolic and hyperbolic stochastic partial differential equations. Nonlinearity **18**, 747–767 (2005)
28. Carlson, D.E.: Linear thermoelasticity. Handbuch der Physik **VI a/2**, 297–346 (1972)
29. Pereira, D.C., Menzala, G.P.: Exponential decay of solutions to a coupled system of equations of linear thermoelasticity. Mat. Appl. Comput. **8**, 193–204 (1989)
30. Cavalcanti, M.M., Guesmia, A.: General decay rates of solutions to a nonlinear wave equation with boundary condition with memory type. Differ. Integr. Equ. **18**, 583–600 (2005)
31. Cavalcanti, M.M., Cavalcanti, V.N.D., Martinez, P.: General decay rate estimates for viscoelastic dissipative systems. Nonlinear Anal. TMA **68**(1), 177–193 (2008)
32. Cavalcanti, M.M., Cavalcanti, V.N.D.: Falcao Nascimento, F.A., Lasiecka, I., Rodrigues, J.H.: Uniform decay rates for the energy of Timoshenko system with the arbitrary speeds of propagation and localized nonlinear damping. Z. Angew. Math. Phys. **65**, 1189–1206 (2014)
33. Cazenave, T.: Semilinear Schrödinger Equations. American Mathematical Society, Providence (2003)
34. Chandrasekharaiah, D.S.: One-dimensional wave propagation in the linear theory of thermoelasticity without energy dissipation. J. Therm. Stress. **19**(8), 695–710 (1996)
35. Chandrasekharaiah, D.S.: A note on the uniqueness of solution in the linear theory of thermoelasticity without energy dissipation. J. Elast. **43**, 279–283 (1996)
36. Chen, G.: Energy decay estimates and exact boundary value controllability for the wave equation in a bounded domain. J. Math. Pures Appl. **58**(3), 249–273 (1979)
37. Chen, Z., Hoffmann, K.H.: On a one-dimensional nonlinear thermoviscoelastic model for structural phase transitions in shape memory alloys. J. Differ. Equ. **112**, 325–350 (1994)
38. Chen, S., Liu, K., Liu, Z.: Spectrum and stability for elastic systems with global or local Kelvin-Voignt damping. SIAM J. Appl. Math. **59**, 651–668 (1998)
39. Chueshov, I., Lasiecka, I.: Global attractors for Mindlin-Timoshenko plates and for their Kirchhoff limits. Milan J. Math. **74**, 117–138 (2006)
40. Chueshov, I.: Introduction to the Theory of Infinite-Dimensional Dissipative Systems. ACTA Scientific Publishing House, Kharkiv, Ukraine (1999). (Translated by Constantin, I. Chueshov from the Russian edition, Translation edited by Khorolska, M.B: (2002))
41. Chueshov, I., Lasiecka, I.: Attractors and long time behavior of von Karman thermoelastic plates. Appl. Math. Optim. **58**, 195–241 (2008)
42. Chueshov, I., Lasiecka, I.: Von Karman Evolution Equations. Springer, New York (2010)
43. Clark, M.R., Guardia, R.R.: On a nonlinear thermoelastic system with nonlocal coefficients. J. Math. Anal. Appl. **433**, 338–354 (2016)
44. Clark, H.R., Clark, M.R., Louredo, A.T., Oliveira, A.M.: A nonlinear thermoelastic system with nonlinear boundary conditions. J. Evol. Equ. **15**, 895–911 (2015)
45. Coleman, B.D., Hrusa, W.J., Owen, D.R.: Stability of equilibrium for a nonlinear hyperbolic system describing heat propagation by second sound in solids. Arch. Ration. Mech. Anal. **94**(3), 267–289 (1986)

46. Copetti, M.I.M., Fernandez, J.R.: A dynamic contact problem in thermoviscoelasticity with two temperatures. Appl. Numer. Math. **77**, 55–71 (2014)
47. Courant, R., Hilbert, D.: Methods of Mathematical Physics, vol. 1. Intersciences, New York (1953)
48. Dafermos, C.M.: On the existence and the asymptotic stability of solution to the equations of linear thermoelasticity. Arch. Ration. Mech. Anal. **29**(4), 241–271 (1968)
49. Dafermos, C.M.: Asymptotic stability in viscoelasticity. Arch. Ration. Mech. Anal. **37**(4), 297–308 (1970)
50. Dafermos, C.M.: Conservation Laws with Dissipation. In: Lakshmilantham, V. (ed.) Nonlinear Phenomena in Mathematical Sciences. Academic Press, New York (1981)
51. Dafermos, C.M.: Global smooth solutions to the initial boundary value problem for the equations of one-dimensional nonlinear thermoviscoelasticity. SIAM J. Math. Anal. **13**(3), 397–408 (1982)
52. Dafermos, C.M.: An abstract Volterra equation with applications to linear viscoelasticity. J. Differ. Equ. **7**, 554–569 (1990)
53. Dafermos, C.M., Hsiao, L.: Global smooth thermomechanical processes in one-dimensional nonlinear thermoelasticity. Nonlinear Anal. TMA **6**(5), 435–454 (1982)
54. Day, A.: Heat Conduction with Linear Thermoelasticity. Springer, New York (1985)
55. Dell'Oro, F.: Asymptotic stability of thermoelastic systems of Bresse type. J. Differ. Equ. **258**, 3902–3927 (2015)
56. Dell'Oro, F., Pata, V.: Lack of exponential stability in Timoshenko systems with flat memory kernels. Appl. Math. Optim. **71**, 79–93 (2015)
57. Desch, W., Miller, R.K.: Exponential stabilization of Volterra integrodifferential equations in Hilbert space. J. Differ. Equ. **70**(3), 366–389 (1987)
58. Desch, W., Miller, R.K.: Exponential stabilization of Volterra integral equations with singular kernels. J. Integr. Equ. Appl. **1**(3), 397–433 (1998)
59. DiPerna, R.J., Lions, P.L.: Ordinary differential equations, transport theory and Sobolev spaces. Invent. Math. **98**, 511–547 (1989)
60. Ducomet, B.: Global existence for a simplified model of nuclear fluid in one dimension. J. Math. Fluid Mech. **2**(1), 1–15 (2000)
61. Ducomet, B.: Simplified models of quantum fluids in nuclear physics. Proc. Partial Differ. Equ. Appl. Math. Bohem. **126**(2), 323–336 (2001)
62. Ducomet, B.: Asymptotic behaviour for a non-monotone fluid in one-dimension: the positive temperature case. Math. Appl. Sci. **24**(8), 543–559 (2001)
63. Elliott, C.M., Tang, Q.: A dynamic contact problem in thermoelasticity. Nonlinear Anal. TMA **23**, 883–898 (1994)
64. El-Karamany, A.S., Ezzat, M.A.: Two-temperature Green-Naghdi theory of type III in linear thermoviscoelastic anisotropic solid. Appl. Math. Model. **39**, 2155–2171 (2015)
65. Evans, L.C.: Partial Differential Equations. Graduate Studies in Mathematics, vol. 19. American Mathematical Society, Providence (1998)
66. Evans, L.C.: Weak Convergence Methods for Nonlinear Partial Differential Equations. CBMS Regional Conference Series in Mathematics, vol. 74. American Mathematical Society, Providence (1990)
67. Ezzat, M.A., El-Karamany, A.S., El-Bary, A.A.: Thermo-viscoelastic materials with fractional relaxation operators. Appl. Math. Model. **39**, 7499–7512 (2015)
68. Fabiano, R.H., Ito, K.: Semigroup theory and numerical approximation for equations arising in linear viscoelascity. SIAM J. Math. Anal. **21**, 374–393 (1990)
69. Fabiano, R.H., Ito, K.: An approximation framework for equations in linear viscoelasticity with strongly singular kernels. Q. Appl. Math. **52**(1), 65–81 (1994)
70. Fabrizio, M., Morro, A.: A boundary condition with memory in electromagnetism. Arch. Ration. Mech. Anal. **136**, 359–381 (1996)
71. Fastovska, T.: Upper semicontinuous attractor for 2D Mindlin-Timoshenko thermoelastic model with memory. Commun. Pure Appl. Anal. **6**, 83–101 (2007)

72. Fastovska, T.: Upper semicontinuous attractor for 2D Mindlin-Timoshenko thermoviscoelastic model with memory. Nonlinear Anal. TMA **71**, 4833–4851 (2009)
73. Fastovska, T.: On the long-time behavior of the thermoelastic plates with second sound. J. Math. Phys. Anal. Geom. **9**(2), 191–206 (2013)
74. Fatori, L.H., JorgeSilva, M.A., Ma, T.F., Yang, Z.: Long-time behavior of a class of thermoelastic plates with nonlinear strain. J. Differ. Equ. **259**, 4831–4862 (2015)
75. Fatori, L.H.: Muñoz Rivera, J.E.: Energy decay for hyperbolic thermoelastic systems of memory type. Q. Appl. Math. **59**(3), 441–458 (2001)
76. Rates of decay to weak thermoelastic Bresse system: Fatori, L.H., Muñoz Rivera, J.E. IMA J. Appl. Math. **75**, 881–904 (2010)
77. The Timoshenko system with history and Cattaneo law: Fatori, L.H., Monteiro, R.N., Ferández Sare, H.D. Appl. Math. Comput. **228**, 128–140 (2014)
78. Feifeisl, E.: Dynamics of Viscous Compressible Fluids. Oxford Univ. Press, Oxford (2004)
79. Feitosa, A.J.R., Oliveiraa, M.L., Miranda, M.M.: Nonlinear boundary stabilization for<pg> </pg> Timoshenko beam system. J. Math. Anal. Appl. **428**, 194–216 (2015)
80. Feng, B., Pelicer, M.L.: Global existence and exponential stability for a nonlinear Timoshenko system with delay. Bound. Value Probl. **206**, 1–13 (2015)
81. Sare, Fernández: H.D., Muñoz Rivera, J.E.: Stability of Timoshenko systems with past history. J. Math. Anal. Appl. **339**(1), 482–502 (2008)
82. Sare, Ferández: H.D., Racke, R.: On the stability of damped Timoshenko systems: Cattaneo versus Fourier's law. Arch. Ration. Mech. Anal. **194**(1), 221–251 (2009)
83. Gao, H.: Muñoz Rivera, J.E.: On the exponential stability of thermoelastic problem with memory. Appl. Anal. **78**, 379–403 (2001)
84. Gawinecki, J.: $L^p - L^q$ time decay estimate for solution of the Cauchy problem for hyperbolic partial differential equations of linear thermoelasticity. Ann. Pol. Math. **2**, 135–145 (1991)
85. Gibson, J.S., Rosen, I.G., Tao, G.: Approximation in control of thermoelastic systems. J. Control Optim. **30**, 1163–1189 (1992)
86. Gilbarg, D., Trudinger, N.: Elliptic Partial Differential Equations of Second Order. Springer, Berlin (1983)
87. Giorgi, C., Vegni, F.M.: The longtime behavior of a nonlinear Reissner-Mindlin plate with exponentially decreasing memory kernels. J. Math. Anal. Appl. **326**, 754–771 (2007)
88. Grasselli, M.: Muñoz Rivera, J.E., Pata, V.: On the energy decay of the linear thermoelastic plate with memory. J. Math. Anal. Appl. **309**, 1–14 (2005)
89. Grasselli, M., Pata, V., Prouse, G.: Longtime behavior of a viscoelastic Timoshenko beam. Discrete Contin. Dyn. Syst. **10**(1–2), 337–348 (2004)
90. Green, A.E., Naghdi, P.M.: A re-examination of the basic postulates of thermomechanics. Proc. R. Soc. Lond. Ser. A **432**(1885), 171–194 (1991)
91. Green, A.E., Naghdi, P.M.: On undamped heat waves in an elastic solid. J. Therm. Stress. **15**(2), 253–264 (1992)
92. Green, A.E., Naghdi, P.M.: On thermoelasticity without energy dissipation. J. Elast. **31**(3), 189–208 (1993)
93. Green, A.E., Naghdi, P.M.: A unified procedure for construction of theories of deformable media, I. Classical continuum physics, II. Generalized continua, III. Mixtures of interacting continua. Proc. R. Soc. Lond. Ser A, **448**, 335–356, 357–377, 379–388 (1995)
94. Dalsen, M.G.V.: Uniform stability for the Timoshenko beam with tip load. J. Math. Anal. Appl. **361**, 392–400 (2010)
95. Dalsen, M.G.V.: Strong stabilization of models incorporating the thermoelastic Reissner-Mindlin plate equations with second sound. Appl. Anal. **90**(9), 1419–1449 (2011)
96. Dalsen, M.G.V.: Stabilization of a thermoelastic Mindlin-Timoshenko plate model revisited. Z. Angew. Math. Phys. **64**, 1305–1325 (2013)
97. Dalsen, M.G.V.: Polynomial decay rate of a thermoelastic MindlinCTimoshenko plate model with Dirichlet boundary conditions. Z. Angew. Math. Phys. **66**, 113–128 (2015)
98. Gronwall, T.H.: Note on the derivatives with respect to a parameter of the solutions of a system of differential equation. Ann. Math. **20**, 292–296 (1919)

99. Guesmia, A., Messaoudi, S.A.: On the control of solutions of a viscoelastic equation. Appl. Math. Comput. **206**(2), 589–597 (2008)
100. Guesmia, A., Messaoudi, S.A.: Some stability resutls for Timoshenko systems with cooperative frictional and infinite memory dampings in the displacement. Acta. Math. Sci. **36B**(1), 1–33 (2016)
101. Guo, B., Zhu, P.: Global existence of smooth solutions to non-linear thermoviscoelastic system with clamped boundary conditions in solid-like materials. Commun. Math. Phy. **203**(2), 365–383 (1999)
102. Gwiazda, P., Klawe, F.Z., Gwiazda, A.S.: Thermo-visco-elasticity for Norton-Hoff-type models. Nonlinear Anal. RWA **26**, 199–228 (2015)
103. Hale, J.: Asymptotic behavior and dynamics in infinite dimensions. In: Hale, J., Martinez-Amores, P. (eds.) Nonlinear Differential Equations. Pitman, Boston (1985)
104. Hale, J.K.: Asymptotic Behaviour of Dissipative Systems, Mathematical Surveys and Monographs, vol. 25. American Mathematical Society, Providence (1988)
105. Hannsgen, K.B., Renardy, Y., Wheeler, R.L.: Effectiveness and robustness with respect to time delays of boundary feedback stabilization in one-dimensional viscoelasticity. SIAM J. Control Optim. **26**(5), 1200–1233 (1988)
106. Hannsgen, K.B., Wheeler, R.L.: Viscoelastic and boundary feedback damping: Precise energy decay rates when creep modes are dominant. J. Integr. Equ. Appl. **2**, 495–527 (1990)
107. Hansen, S.W.: Exponential energy decay in a linear thermoelastic rod. J. Math. Anal. Appl. **167**(2), 429–442 (1992)
108. Hardy, G.H., Littlewood, J.E., Pólya, G.: Inequalities, 2nd edn. Cambridge University Press, Cambridge (1952)
109. Hasse, R.W.: Approaches to nuclear friction. Rep. Prog. Phys. **41**(7), 1027–1101 (1978)
110. Henry, D.: Geometric Theory of Semilinear Parabolic Equations. Lecture Notes in Mathematical, vol. 840. Springer, New York (1981)
111. Henry, D., Lopes, O., Perissinotto, A.: On the essential spectrum of a semigroup of thermoelasticity. Nonlinear Anal. TMA **21**(1), 65–75 (1993)
112. Hille, E., Phillips, R.S.: Funct. Anal. Semigroups. American Mathematical Society, Providence (1957)
113. Hoffimann, K.H., Zochowski, A.: Existence of solutions to some non-linear thermoelastic systems with viscosity. Math. Meth. Appl. Sci. **15**, 187–204 (1992)
114. Hölder, O.: Ü ber einen Mittelworthssatz. Nachr. Ges. Wiss: GoÖttingen **38–47**, (1889)
115. Hrusa, W.J., Messaoudi, S.: On formation of singularities in one-dimensional nonlinear thermoelasticity. Arch. Ration. Mech. Anal. **111**, 135–151 (1990)
116. Hrusa, W.J., Tarabek, M.A.: On smooth solutions of the Cauchy problem in one-dimensional nonlinear thermoelasticity. Q. Appl. Math. **47**, 631–644 (1989)
117. Hsiao, L., Jian, H.: Asymptotic behaviour of solutions to the system of one-dimensional nonlinear themoviscoelasticity. Chin. Ann. Math. **19B**(2), 143–152 (1998)
118. Hsiao, L., Luo, T.: Large time behaviour of solutions to the equations of one-dimensional nonlinear thermoviscoelasticity. Q. Appl. Math. **56**(2), 201–219 (1998)
119. Huang, F.: Characteristic condition for exponential stability of linear dynamical systems in Hilbert spaces. Ann. Differ. Equ. **1**, 43–56 (1985)
120. Jiang, S.: Global large solutions to initial boundary value problems in one-dimensional thermoviscoelasticity. Q. Appl. Math. **51**(4), 731–744 (1993)
121. Jiang, S.: Exponential decay and global existence of spherically symmetric solutions in thermoelasticity. Chin. Ann. Math. **19A**(5), 629–640 (1998)
122. Jiang, S.: Global smooth solutions to a one-dimensional nonlinear thermoviscoelastic model. Adv. Math. Sci. Appl. **7**, 771–787 (1997)
123. Jiang, S., Racke, R.: Evolution Equations in Thermoelasticity. Pitman Series Monographs and Surveys in Pure and Applied Mathematics, vol. 112. Chapman & Hall/CRC, Boca Raton (2000)
124. Jiang, S., Muñoz Rivera, J.E., Racke, R.: Asymptotic stability and global existence in thermoelasticity with symmetry. Q. Appl. Math. **56**(2), 259–275 (1998)

125. Kato, T.: Nonlinear semigroups and evolution equations. J. Math. Soc. Jpn. **19**, 508–520 (1967)
126. Kato, T.: Linear evolution equation of hyperbolic type II. J. Math. Soc. Jpn. **25**, 648–666 (1973)
127. Kim, J.U.: Global existence of solutions of the equations of one-dimensional thermoviscoelasticity with initial data in BV and L^1. Ann. Scoula Norm. Sup. Pisa **10**, 357–427 (1983)
128. Kim, J.U.: On the energy decay of a linear thermoelastic bar and plate. SIAM J. Math. Anal. **23**, 889–899 (1992)
129. Kim, J.U., Renardy, Y.: Boundary control of the Timoshenko beam. SIAM J. Control Optim. **25**(6), 1417–1429 (1987)
130. Komornik, V., Zuazua, E.: A direct method for the boundary stabilization of the wave equation. J. Math. Pures Appl. **69**(1), 33–54 (1990)
131. Krasnosel'skii, M.A., Vainikko, G.M., Zabreiko, P.P., Ruttskii, Y.B., Stetesenko, V.Y.: Approximate solution of operator equations. Noordho, Groningen (1972)
132. Krejci, P., Sprekels, J.: On a system of nonlinear PDEs with temperature-dependent hysteresis in one-dimensional thermoplasticity. J. Math. Anal. Appl. **209**(1), 25–46 (1997)
133. Krejci, P., Sprekels, J.: Weak stabilization of solutions to PDEs with hysteresis in thermoviscoelastoplasticity. In: Agawal, R.P., Meuman, F., Vosmansky, J. (eds.) EQUADIFF 9-Proceedings, pp. 81–96. Masaryk Univ, Brno (1998)
134. Krejci, P., Sprekels, J.: Temperature-dependent hysteresis in one-dimensional thermoviscoelastoplasticity. Appl. Math. **43**(3), 173–205 (1998)
135. Kupradze, V.D.: Three-dimensional problems of the mathematical theory of elasticity and thermoelasticity. North-Holland, Amsterdam (1979)
136. Lagnese, J.: Boundary stabilization of linear elastodynamic systems. SIAM J. Control Optim. **21**(6), 968–984 (1983)
137. Lagnese, J.E., Lions, J.L.: Modelling Analysis and Control of Thin Plates. Masson, Paris (1988)
138. Lazzari, B., Nibbi, R.: On the energy decay of a linear hyperbolic thermoelastic system with dissipative boundary. J. Therm. Stress. **30**, 1–14 (2007)
139. Lasiecka, I.: Mathematical Control Theory of Coupled PDEs. CBMS-NSF Lecture Notes. SIAM, Philadelphia (2002)
140. Lasiecka, I., Triggiani, R.: Control Theory for Partial Differential Equations. Cambridge University Press, Cambridge (2000)
141. Lazzari, B., Nibbi, R.: On the exponential decay in thermoelasticity without energy dissipation and of type III in presence of an absorbing boundary. J. Math. Anal. Appl. **338**(1), 317–329 (2008)
142. Leugering, G.: On boundary feedback stabilization of a viscoelastic membrance. Dyn. Stab. Syst. **4**(1), 71–79 (1989)
143. Leugering, G.: On boundary feedback stabilization of a viscoelastic beam. Proc. R. Soc. Edinb. Sect. A **114**, 57–69 (1990)
144. Lebeau, G., Zuazua, E.: Decay rates for the three-dimensional linear system of thermoelasticity. Arch. Ration. Mech. Anal. **148**(3), 179–231 (1999)
145. Lieb, E., Loss, M.: Analysis. Graduate Studies in Mathematics, vol. 14, 2nd edn. American Mathematical Society, Providence (2001)
146. Linz, P.: Analytical and Numerical Methods for Volterra Equations. SIAM Studies in Applied Mathematics. SIAM, Philadephia (1985)
147. Liu, K., Liu, Z.: On the type of C_0-semigroup associated with the abstract linear viscoelastic system. Z. Angew. Math. Phys. **47**(1), 1–15 (1996)
148. Liu, W.: Partial exact controllability and exponential stability in higher dimensional linear thermoelascity. ESAIM: Control Optim. Calc. Var. **3**, 23–48 (1998)
149. Liu, W.: The exponential stabilization of higher-dimensional linear system of thermoviscoelasticity. J. Math. Pures Appl. **77**(4), 355–386 (1998)
150. Liu, Z., Liu, K., Rao, B.: Exponential stability of an abstract nondissipative linear system. SIAM J. Control Optim. **40**, 149–165 (2001)

151. Liu, Z., Rao, B.: Characteristic of polynomial decay rate for the solution of linear evolution equation. Z. Angew. Math. Phys. **56**, 630–644 (2005)
152. Liu, Z., Rao, B.: Energy decay rate of the thermoelastic Bresse system. Z. Math. Phys. **60**(1), 54–69 (2009)
153. Liu, Z., Quintanilla, R.: Energy deday rate of a mixed type II and type III thermoelastic system. Discrete Contin. Dyn. Syst. Ser. B **14**(4), 1433–1444 (2010)
154. Liu, Z., Zheng, S.: Exponential stability of the semigroup associated a thermoelastic system. Q. Appl. Math. **51**, 535–545 (1993)
155. Liu, Z., Zheng, S.: On the exponential stability of linear viscoelasticity and thermoviscoelasticity. Q. Appl. Math. **54**(1), 21–31 (1996)
156. Liu, Z., Zheng, S.: Semigroups Associated with Dissipative Systems. Chapman and Hall/CRC, Boca Raton (1999)
157. Lord, H., Shulman, Y.: A generalized dynamical theory of thermoelasticity. J. Mech. Phys. Solids **15**, 299–309 (1967)
158. Ma, Z.: Exponential stability and global attractors for a thermoelastic Bresse system. Adv. Differ. Equ. **1**, 1–17 (2010)
159. Ma, Z.: Global attractors for thermoelastic plates with memory. J. Shanghai Second Polytech. Univ. **31**(4), 316–319 (2014)
160. Ma, Z.: Polynomial stability for Timoshenko-type system with past history. Appl. Mech. Mater. **623**, 78–84 (2014)
161. Ma, Z., Qin, Y.: Polynomial stability for a Timoshenko-type system with a past history. Preprint
162. Ma, Z., Qin, Y.: Global attractors for thermoelastic plates with memory. Preprint (2012)
163. Ma, Z., Zhang, L., Yang, X.: Exponential stability for a Timoshenko-type system with history. J. Math. Anal. Appl. **380**, 299–312 (2011)
164. Martinez, P.: A new method to obtain decay rate estimates for dissipative systems with localized damping. Rev. Mat. Comput. **12**(1), 419–444 (1999)
165. Messaoudi, S.A.: On weak solutions of semilinear thermoelastic equations. Magreb Math. Rev. **1**(1), 31–40 (1992)
166. Messaoudi, S.A.: A blow up result in a multidimensional semilinear thermoelastic system. Electron. J. Differ. Equ. **2001**(30), 1–9 (2001)
167. Messaoudi, S.A.: Local existence and blow up in thermoelasticity with second sound. Commun. Partial Differ. Equ. **26**(8), 1681–1693 (2002)
168. Messaoudi, S.A.: Decay of solutions of a nonlinear hyperbolic system describing heat propagation by second sound. Appl. Anal. **81**(2), 201–209 (2002)
169. Messaoudi, S.A.: General decay of solution energy in a viscoelastic equation with a nonlinear source. Nonlinear Anal. TMA **69**, 2589–2598 (2008)
170. Messaoudi, S.A., Houari, B.S.: Exponential stability in one-dimensional nonlinear thermoelasticity with second sound. Math. Mech. Appl. Sci. **28**(2), 205–232 (2005)
171. Messaoudi, S.A., Said-Houari, B.: Energy decay in a Timoshenko-type system with history in thermoelasticity of type III. Adv. Diff. Equ. **14**, 375–400 (2009)
172. Messaoudi, S.A., Mustafa, M.I.: On the stabilization of the Timoshenko system by a weak nonlinear dissipation. Math. Meth. Appl. Sci. **32**(4), 454–469 (2009)
173. Messaoudi, S.A., Mustafa, M.I.: A stability result in a memory-type Timoshenko system. Dyn. Syst. Appl. **2**, 957–972 (2013)
174. Messaoudi, S.A., Mustafa, M.I.: On the control of solutions of viscoelastic equations with boundary feedback. Nonlinear Anal. RWA **10**(5), 3132–3140 (2009)
175. Messaoudi, S.A., Pokojovy, M., Said-Houari, B.: Nonlinear damped Timoshenko systems with second sound-Global existence and exponential stability. Math. Meth. Appl. Sci. **32**(5), 505–534 (2009)
176. Messaoudi, S.A., Said-Houari, B.: Energy decay in a Timoshenko-type system of thermoelasticity of type III. J. Math. Anal. Appl. **348**(1), 298–307 (2008)
177. Messaoudi, S.A., Said-Houari, B.: Uniform decay in a Timoshenko-type system with past history. J. Math. Anal. Appl. **360**(2), 459–475 (2009)

178. Messaoudi, S.A., Said-Houari, B.: Energy decay in a transmission problem in thermoelasticity of type III. IMA J. Appl. Math. **74**, 344–360 (2009)
179. Messaoudi, S.A., Tatar, N.: Global existence and uniform stability of solutions for a quasilinear viscoelastic problem. Math. Meth. Appl. Sci. **30**(6), 665–680 (2007)
180. Messaoudi, S.A., Tatar, N.: Exponential and polynomial decay for a quasilinear viscoelastic problem. Nonlinear Anal. TMA **68**(4), 785–793 (2008)
181. Meyvaci, M.: Structural stability for a linear system of thermoelastic plate equations. Appl. Math. Comput. **242**, 664–668 (2014)
182. Miao, C.: Harmonic Analysis and its Applications in Partial Differential Equations, 2nd edn. Science Press, Beijing (2004). (in Chinese)
183. Miao, C.: Modern Methods for Nonlinear Wave Equations. Lectures in Contemporary Mathematics, vol. 2. Science Press, Beijing (2005). (in Chinese)
184. Muñoz Rivera, J.E.: Energy decay rate in linear thermoelascity. Funkcial Ekvac. **35**, 19–30 (1992)
185. Muñoz Rivera, J.E.: Decomposition of the displacement vector field and decay rates in linear thermoelasticity. SIAM J. Math. Anal. **24**(2), 390–406 (1993)
186. Muñoz Rivera, J.E.: Asymptotic behavior in n-dimensional thermoelasticity. Appl. Math. Lett. **10**(5), 47–53 (1997)
187. Muñoz Rivera, J.E., Andrade, D.: Exponential decay of nonlinear wave equation with a viscoelastic boundary condition. Math. Meth. Appl. Sci. **23**, 41–61 (2000)
188. Muñoz Rivera, J.E., Avila, A.I.: Rates of decay to non homogeneous Timoshenko model with tip body. J. Differ. Equ. **258**, 3468–3490 (2015)
189. Muñoz Rivera, J.E., Barreto, R.K.: Decay rates of solutions to thermoviscelastic plates with memory. IMA J. Appl. Math. **60**, 263–283 (1998)
190. Muñoz Rivera, J.E., Qin, Y.: Global existence and exponential stability in one-dimensional nonlinear thermoelasticity with thermal memory. Nonlinear Anal. TMA **51**(1), 11–32 (2002)
191. Muñoz Rivera, J.E., Racke, R.: Global stability for damped Timoshenko systems. Discrete Contin. Dyn. Syst. **9**, 1625–1639 (2003)
192. Muñoz Rivera, J.E., Racke, R.: Mildly dissipative nonlinear Timoshenko systems - Global existence and exponential stability. J. Math. Anal. Appl. **276**(1), 248–278 (2002)
193. Muñoz Rivera, J.E., Racke, R.: Timoshenko systems with indefinite damping. J. Math. Anal. Appl. **341**(2), 1068–1083 (2008)
194. Mustafa, M.I.: On the decay rates for thermoviscoelastic systems of type III. Appl. Math. Comput. **239**(15), 29–37 (2014)
195. Nappa, L.: Spatial decay estimates for the evolution equations of linear thermoelasticity without energy dissipation. J. Therm. Stress. **21**(5), 581–592 (1998)
196. Najdi, N., Wehbe, A.: Weakly locally thermal stabilization of Bresse systems. Electron. J. Differ. Equ. **182**, 1–19 (2014)
197. Navarro, C.B.: Asymptotic stability in linear thermoviscoelasticity. J. Math. Anal. Appl. **65**(2), 399–431 (1978)
198. Nibbi, R., Polidoro, S.: Exponential decay for Maxwell equations with a boundary memory condition. J. Math. Anal. Appl. **302**, 30–55 (2005)
199. Nirenberg, L.: Topics in Nonlinear Functional Analysis, Courant Institute of Mathematical Sciences (1974)
200. Pata, V., Zucchi, A.: Attractors for a damped hyperbolic equation with linear memory. Adv. Math. Sci. Appl. **11**(2), 505–529 (2001)
201. Pazy, A.: Semigroup of Linear Operators and Appplications to Partial Differential Equations. Springer, New York (1983)
202. Pei, P., Rammaha, M.A., Toundykov, D.: Global well-posedness and stability of semilinear Mindlin-Timoshenko system. J. Math. Anal. Appl. **418**, 535–568 (2014)
203. Potomkin, M.: Asymptotic behavior of thermoviscoelastic Berger plate. Commun. Pure Appl. Anal. **9**(1), 161–192 (2010)
204. Propst, G., Prüss, J.: On wave equations with boundary dissipation of memory type. J. Integr. Equ. Appl. **8**, 99–123 (1996)

205. Prüss, J.: On the spectrum of C_0 semigroup. Trans. Am. Math. Soc. **284**, 847–857 (1984)
206. Qin, Y.: Asymptotic behaviour for global smooth solution to a one-dimensional nonlinear thermoviscoelastic system. J. Partial Differ. Equ. **12**(2), 111–134 (1999)
207. Qin, Y.: Global existence and asymptotic behaviour of the solution to the system in one-dimensional nonlinear thermoviscoelasticity. Q. Appl. Math. **59**(1), 113–142 (2001)
208. Qin, Y.: Universal attractor in H^4 for the nonlinear one-dimensional compressible Navier-Stokes equations. J. Differ. Equ. **207**(1), 21–72 (2004)
209. Qin, Y.: Exponential stability and maximal attractors for a one-dimensional nonlinear thermoviscoelasticity. IMA J. Appl. Math. **70**(4), 509–526 (2005)
210. Qin, Y.: Nonlinear Parabolic-Hyperbolic Systems and Their Attractors. Operator Theory, Advances in PDEs, vol. 184. Birkhäser, Basel (2008)
211. Qin, Y., Deng, S., Huang, L., Ma, Z., Su, X.: Global existence for the three-dimensional thermoelastic equations of type II. Q. Appl. Math. **2**, 333–348 (2010)
212. Qin, Y., Fang, J.: Global attractor for a nonlinear thermoviscoelastic model with a non-convex free energy density. Nonlinear Anal. TMA. **65**(4), 892–917 (2006)
213. Qin, Y., Feng, B., Zhang, M.: Uniform attractors for a non-autonomous viscoelastic equation with a past history. Nonlinear Anal. TMA **101**, 1–15 (2014)
214. Qin, Y., Huang, L.: Global Well-posedness of Nonlinear Parabolic-Hyperbolic Coupled Systems. Frontiers in Mathematics. Birkhäuser, Basel (2012)
215. Qin, Y., Liu, X., Wang, T.: Global Existence and Uniqueness of Nonlinear Evolutionary Fluid Equations. Frontiers in Mathematics. Birkhäuser, Basel (2015)
216. Qin, Y.: Lü, Tzon-tzer: Global attractor for a nonlinear viscoelasticity. J. Math. Anal. Appl. **341**(2), 975–997 (2008)
217. Qin, Y., Liu, H., Song, C.: Global attractor for a nonlinear thermoviscoelastic system in shape memory alloys. Proc. R. Soc. Edingb. Sect. A **138**(5), 1103–1135 (2008)
218. Qin, Y., Ma, T., Cavalcanti, M.M., Andrade, D.: Exponential stability in H^4 for the Navier-Stokes equations of viscous and heat-conductive fluid. Commun. Pure Appl. Anal. **4**(3), 635–664 (2005)
219. Qin, Y, Ma, Z.: Energy decay and global attractors for Timoshenko-type system with past history. Preprint (2012)
220. Qin, Y, Ma, Z.: Global existence and exponential stabilization for the higher-dimensional linear thermoelastic system of type III. Preprint (2013)
221. Qin, Y., Ma, Z.: Decay of solutions and global attractors for thermoviscoelastic systems. Acta. Appl. Math. **117**, 195–214 (2012)
222. Qin, Y., Ma, Z.: Huang, Lan: Global existence of solutions in H^4 for a nonlinear thermoviscoelastic equations with non-monotone pressure. J. Elast. **94**, 15–31 (2009)
223. Qin, Y., Ma, Z., Yang, X.: Exponential stability for nonlinear thermoelastic equations with second sound. Nonlinear Anal. RWA **11**, 2502–2513 (2010)
224. Qin, Y., Muñoz Rivera, J.E.: Universal attractors for a nonlinear one-dimensional heat-conductive viscous real gas. Proc. R. Soc. Edinb. Sect. A **132**(3), 685–709 (2002)
225. Qin, Y., Muñoz, J.E.: Exponential stability and universal attractors for the Navier-Stokes equations of compressible fluids between two horizontal parallel plates in $\mathbb{R}^3$. Appl. Numer. Math. **47**(2), 209–235 (2003)
226. Qin, Y., Muñoz Rivera, J.E.: Global existence and exponential stability of solutions to thermoelastic equations of hyperbolic type. J. Elast. **75**(2), 125–145 (2004)
227. Qin, Y., Xu, L., Ma, Z.: Global existence and exponential stability for a thermoelastic equations of type II. J. Zhenzhou Univ., (Nat. Sci. Ed.) **40**, 1–11 (2008)
228. Qin, Y., Yang, X.G., Ma, Z.: Global existence of solutions for the thermoelastic Bresse system. Commu. Pure. Appl. Anal. **13**(4), 1395–1406 (2014)
229. Quintanilla, R.: Structural stability and continuous dependence of solutions in thermoelasticity of type III. Discrete Contin. Dyn. Syst. Ser. B **1**(4), 463–470 (2001)
230. Quintanilla, R.: Existence in thermoelasticity without energy dissipation. J. Therm. Stress. **25**(2), 195–202 (2002)

231. Quintanilla, R.: Convergence and structural stability in thermoelasticity. Appl. Math. Comput. **135**(2–3), 287–300 (2003)
232. Quintanilla, R.: Some remarks on growth and uniqueness in thermoelasticity. Int. J. Math. Sci. **2003**, 617–623 (2003)
233. Quintanilla, R., Racke, R.: Stability in thermoelasticity of type III. Discrete Contin. Dyn Syst. Ser. B **3**(3), 383–400 (2003)
234. Racke, R.: On the Cauchy problem in nonlinear 3-d thermoelasticity. Math. Z. **203**, 649–682 (1990)
235. Racke, R.: Thermoelasticity with second sound-exponential stability in linear and nonlinear 1-d. Math. Meth. Appl. Sci. **25**(5), 409–441 (2002)
236. Racke, R.: Asymptotic behavior of solutions in linear 2- or 3-d thermoelasticity with second sound. Q. Appl. Math. **61**(2), 315–328 (2003)
237. Racke, R., Said-Houari, B.: Global existence and decay property of the Timoshenko system in thermoelasticity with second sound. Nonlinear Anal. TMA **75**, 4957–4973 (2012)
238. Racke, R., Shibata, Y.: Global smooth solutions and asymptotic stability in one-dimensional nonlinear thermoelasticity. Arch. Ration. Mech. Anal. **116**, 1–34 (1991)
239. Racke, R., Wang, Y.: Asymptotic behavior of discontinuous solutions to thermoelastic systems with second sound. Z. Anal. Anwendung **24**(1), 117–135 (2005)
240. Racke, R., Shibata, Y., Zheng, S.: Global solvability and exponential stability in one-dimensional nonlinear thermoelasticity. Q. Appl. Math. **51**(4), 751–763 (1993)
241. Racke, R., Zheng, S.: Global existence and asymptotic behaviour in nonlinear themoviscoelasticity. J. Differ. Equ. **134**(1), 46–67 (1997)
242. Raposo, C.A., Ferreira, J., Santos, M.L., Castro, N.N.O.: Exponential stability for the Timoshenko system with two weak dampings. Appl. Math. Lett. **18**(5), 535–541 (2005)
243. Reissig, M., Wang, Y.: Cauchy problems for linear thermoelastic systems of type III in one space variable. Math. Meth. Appl. Sci. **28**(11), 1359–1381 (2005)
244. Rifo, S., Villagran, O.V.: Muñoz Rivera, J.E.: The lack of exponential stability of the hybrid Bresse system. J. Math. Anal. Appl. **436**(1), 1–15 (2016)
245. Romero, P.G.: Polynomial decay to thermoelastic plates with memory. Q. Appl. Math. **3**, 553–558 (2009)
246. Rudin, W.: Functional Analysis, 2nd edn. McGraw-Hill Book Company, New York (1991)
247. Said-Houari, B., Kasimov, A.: Decay property of Timoshenko system in thermoelasticity. Math. Meth. Appl. Sci. **35**(3), 314–333 (2012)
248. Schwartz, L.: Théorie des distributions, I, II, Act. Sci. Ind., 1091, 1122. Hermann et Cie, Paris (1951)
249. Serrin, J.: Mathematical principles of classical fluids mechanics. Hanbuch der Physik, vol. VIII/ 1, pp. 125–262. Springer, Berlin (1972)
250. Shang, C.: Global attractor for the Ginzburg-Landau thermoviscoelastic systems with hinged boundary conditions. J. Math. Anal. Appl. **343**, 1–21 (2008)
251. Shen, W., Zheng, S., Zhu, P.: Global existence and asymptotic behaviour of weak solutions to nonlinear thermoviscoelastic systems with clamped boundary conditions. Q. Appl. Math. **57**(1), 93–116 (1999)
252. Shi, P., Shillor, M.: Existence of a solution to the n-dimensional problem of thermoelastic contact. Commun. Partial Differ. Equ. **17**, 1597–1618 (1992)
253. Shi, P., Xu, Y.: Decoupling of the quasistatic system of thermoelasticity on the unit disk. J. Elast. **31**, 209–218 (1993)
254. Shi, D., Feng, D.: Exponential decay of Timoshenko beam with lacally distributed feedback. IMA J. Math. Control Inform. **18**(3), 395–403 (2001)
255. Shibata, Y.: Neumann problem for one-dimensional nonlinear thermoelasticity. Commun. Partial Differ. Equ. **27**(2), 457–480 (1992)
256. Shibata, Y.: Global in time existence of smooth solutions of nonlinear thermoviscoelastic equations. Math. Meth. Appl. Sci. **18**, 871–895 (1995)
257. Slemrod, M.: Global existence, uniqueness, and asymptotic stability of classical solutions in one-dimensional thermoelasticity. Arch. Ration. Mech. Anal. **76**(2), 97–133 (1981)

258. Sobolev, S.L.: Partial Differential Equations of Mathematical Physics. International Series of Monographs in Pure and Applied Mathematics, vol. 50. Pergamon Press, Oxford (1964). [II.9]
259. Soufyane, A.: Stabilisation de la poutre de Timoshenko. C. R. Acad. Sci. Paris Ser. **I**(328), 731–734 (1999)
260. Soufyane, A., Wehbe, A.: Uniform stabilization for the Timoshenko beam by a locally distributed damping. Electron. J. Differ. Equ. **2003**(29), 1–14 (2003)
261. Sprekels, J., Zheng, S.: Global solutions to the equations of a Ginzburg-Landau theory for structural phase transition in shape memory alloys. Phys. D **39**, 59–76 (1989)
262. Sprekels, J., Zheng, S.: Maximal attractor for the system of a Landau-Ginzburg theory for structural phase transitions in shape memory alloys. Phys. D **121**, 252–262 (1998)
263. Sprekels, J., Zheng, S., Zhu, P.: Asymptotic behavior of the solutions to a Landau-Ginzburg system with viscosity for martensitic phase transitions in shape memory alloys. SIAM J. Math. Anal. **29**, 69–84 (1998)
264. Staffans, O.J.: On a nonlinear hyperbolic Volterra equation. SIAM J. Math. Anal. **11**(5), 793–812 (1980)
265. Sureau, E.: La matiere nucleaire: desetoiles aux noyaux. Hermann, Paris (1998)
266. Tarabek, M.A.: On the existence of smooth solutions in one-dimensional thermoelasticity with second sound. Q. Appl. Math. **50**(4), 727–742 (1992)
267. Taylor, S.W.: A smoothing property of a hyperbolic system and boundary controllability. J. Comput. Appl. Math. **114**(1), 23–40 (2000)
268. Taylor, M.E.: Partial Differential Equations, I, II, III, 2nd edn. Springer, New York (2010)
269. Temam, R.: Infinite-Dimensional Dynamical Systems in Mechanics and Physics. Springer, New York (1988)
270. Timoshenko, S.: On the correction for shear of the differential equation for transverse vibrations of prismaticbars. Philos. Mag. **41**, 744–746 (1921)
271. Triebel, H.: Interpolation Theory. Function Spaces. Differential Operators. North-Holland Publishing Company, Amsterdam (1978)
272. Volokh, K.Y.: Non-linear thermoelasticity with energy limiters. Int. J. Non-Linear Mech. **76**, 169–175 (2015)
273. Walter, W.: Differential and Integral Inequalities. Springer, Berlin (1964)
274. Wang, J., Guo, B.: On dynamic behavior of hyperbolic system derived from a thermoelastic equation with memory type. J. Frankl. Inst. **344**(2), 75–96 (2007)
275. Wang, W., Wang, Z.: The pointwise estimates to solutions for one-dimensional linear thermoelastic system with second sound. J. Math. Anal. Appl. **326**(2), 1061–1075 (2007)
276. Watson, S.J.: Unique global solvability for initial-boundary value problems in one-dimensional nonlinear thermoviscoelasticity. Arch. Ration. Mech. Anal. **153**(1), 1–37 (2000)
277. Watson, S.J.: A priori bounds in one-dimensional nonlinear thermoviscoelasticity. Contemp. Math. **225**, 229–238 (2000)
278. Wu, Z., Xu, J.: Global existence in H^2 for a one-dimensional non-monotone fluid. Chin. Q. J. Math. **22**(2), 296–304 (2007)
279. Xu, J., Mori, N., Kawashima, S.: Global existence and minimal decay regularity for the Timoshenko system: the case of non-equal wave speeds. j. differ. equ. **259**, 5533–5553 (2015)
280. Yan, Q.: Boundary stabilization of Timoshenko beam. Syst. Sci. Math. Sci. **13**(4), 376–384 (2000)
281. Yang, L., Wang, Y.: Propagation of singularities in Cauchy problems for quasilinear thermoelastic systems in three space variables. J. Math. Anal. Appl. **291**(2), 638–652 (2004)
282. Yang, L., Wang, Y.: Well-posedness and decay estimates for Cauchy problems of linear thermoelastic systems of type III in 3-D. Indiana Univ. Math. J. **55**, 1333–1362 (2006)
283. Yosida, K.: Functional Analysis, 6th edn. Springer, Berlin (1980)
284. Young, W.H.: On classes of summable functions and their Fourier series. Proc. R. Soc. Lond. A **87**, 225–229 (1912)
285. Zhang, X., Zuazua, E.: Decay of solutions of the system of thermoelasticity of type III. Commun. Contemp. Math. **5**(1), 1–59 (2003)

286. Zheng, S.: Nonlinear Parabolic Equations and Hyperbolic-Parabolic Coupled Systems. Pitman Series Monographs and Surveys in Pure and Applied Mathematics, vol. 76. Longman Group Limitted, London (1995)
287. Zheng, S.: Nonlinear Evolution Equations. Pitman Monographs and Surveys in Pure and Applied Mathematics, vol. 133. CRC Press, USA (2004)
288. Zheng, S., Shen, W.: Global solutions to the Cauchy problem of the equations of one-dimensional thermoviscoelasticity. J. Partial Differ. Equ. **2**, 26–38 (1989)
289. Zheng, S., Shen, W.: Global solutions to the Cauchy problem of quasilinear hyperbolic-parabolic coupled systems. Sci. Sinica A **3**(11), 1133–1149 (1987)

Index

Y. Qin and Z. Ma, *Global Well-posedness and Asymptotic Behavior of the Solutions to Non-classical Thermo(visco)elastic Models*,
DOI 10.1007/978-981-10-1714-8

CPSIA information can be obtained
at www.ICGtesting.com
Printed in the USA
LVOW01*2058230816
501523LV00012B/159/P

9 789811 017131